Praise for Martin Gardner and *Are Universes Thicker th*

"This Renaissance man of w ch a national treasure as ever, whet epts, debunking the paranormal or gotten writers and books. . . . Something about Ga s prose—straight ahead, factual, free of literary pretension—is deliciously addictive. . . . It's good to be in periodic touch with a mind of such rare sense and clarity."

—Michael Dirda, *Washington Post Book World*

"Gardner is a living national treasure and a phenomenon of mind and nature. . . . The pieces [in *Are Universes Thicker than Blackberries?*] ring loud with clarity of expression, breadth and depth of recall and crackling energy." —Michael Pakenham, *Baltimore Sun*

"Martin Gardner's contribution to contemporary intellectual culture is unique—in its range, its insight, and its understanding of hard questions that matter." —Noam Chomsky

"Gardner's books have all been informed by the sound understanding of science, mathematics, and philosophy. . . . Those disciplines have given him a firm footing for the center of his life's work, the exposure of fraud in charlatans pretending to be spiritualists, religious freaks posing as scientists, and scientists thinking they are philosophers."

—James Franklin, *The New Criterion*

"Martin Gardner is one of our most awe inspiring writers. . . . He is irreplaceable—at least in this universe." —Eric Wolff, *New York Sun*

"[Martin Gardner] has been patrolling the boundary between science and pseudoscience for more than a half century. . . . Perhaps it is too much to ask that every page of a book offer both instruction and pleasure. Of Gardner's books, I think it can very nearly be said that every blessed page offers, if not both, at least one." —John Derbyshire, *Washington Times*

Books by Martin Gardner

Fads and Fallacies in the Name of Science
Mathematics, Magic, and Mystery
Great Essays in Science (ed.)
Logic Machines and Diagrams
The Scientific American Book of Mathematical Puzzles and Diversions
The Annotated Alice
The Second Scientific American Book of Mathematical Puzzles and Diversions
Relativity for the Millions
The Annotated Snark
The Ambidextrous Universe
The Annotated Ancient Mariner
New Mathematical Diversions from Scientific American
The Annotated Casey at the Bat
Perplexing Puzzles and Tantalizing Teasers
The Unexpected Hanging and Other Mathematical Diversions
Never Make Fun of a Turtle, My Son (verse)
The Sixth Book of Mathematical Games from Scientific American
Codes, Ciphers, and Secret Writing
Space Puzzles
The Snark Puzzle Book
The Flight of Peter Fromm (novel)
Mathematical Magic Show
More Perplexing Puzzles and Tantalizing Teasers
The Encyclopedia of Impromptu Magic
Aha! Insight
Mathematical Carnival
Science: Good, Bad, and Bogus
Science Fiction Puzzle Tales
Aha! Gotcha
Wheels, Life, and Other Mathematical Amusements
Order and Surprise
The Whys of a Philosophical Scrivener
Puzzles from Other Worlds
The Magic Numbers of Dr. Matrix

Knotted Doughnuts and Other Mathematical Entertainments

The Wreck of the Titanic Foretold?

Riddles of the Sphinx

The Annotated Innocence of Father Brown

The No-Sided Professor (short stories)

Time Travel and Other Mathematical Bewilderments

The New Age: Notes of a Fringe Watcher

Gardner's Whys and Wherefores

Penrose Tiles to Trapdoor Ciphers

How Not to Test a Psychic

The New Ambidextrous Universe

More Annotated Alice

The Annotated Night Before Christmas

Best Remembered Poems (ed.)

Fractal Music, Hypercards, and More

The Healing Revelations of Mary Baker Eddy

Martin Gardner Presents

My Best Mathematical and Logic Puzzles

Classic Brainteasers

Famous Poems of Bygone Days (ed.)

Urantia: The Great Cult Mystery

The Universe Inside a Handkerchief

On the Wild Side

Weird Water and Fuzzy Logic

The Night Is Large

Mental Magic

The Annotated Thursday

Visitors from Oz (novel)

The Annotated Alice: The Definitive Edition

Did Adam and Eve Have Navels?

The Colossal Book of Mathematics

Are Universes Thicker than Blackberries?

Discourses on Gödel, Magic Hexagrams, Little Red Riding Hood, and other Mathematical and Pseudoscientific Topics

by Martin Gardner

W. W. Norton & Company
NEW YORK LONDON

Printed in the United States of America
First published as a Norton paperback 2004

Manufacturing by Quebecor Fairfield
Book design by Brooke Koven
Production manager: Julia Druskin

Library of Congress Cataloging-in-Publication Data

Gardner, Martin, 1914–
Are universes thicker than blackberries? : discourses on Gödel, magic hexagrams, Little Red Riding Hood, and other mathematical and pseudoscientific topics / by Martin Gardner.— 1st ed.
p. cm.
Includes index.
ISBN 0-393-05742-9 (hardcover)
1. Mathematics—Miscellanea. 2. Science—Miscellanea. 3. Pseudoscience.
I. Title.

QA99.G37 2003
510—dc21

2003001412

ISBN 0-393-32572-5 pbk.

W. W. Norton & Company, Inc.
500 Fifth Avenue, New York, N.Y. 10110
www.wwnorton.com

W. W. Norton & Company Ltd.
Castle House, 75/76 Wells Street, London W1T 3QT

2 3 4 5 6 7 8 9 0

To Dana Richards, friend, mathematician, Sherlockian, puzzle buff, and bibliographer extraordinary.

CONTENTS

PART IV. LITERATURE

PART V. MOONSHINE

PREFACE

This book contains the last of several collections of my column "Notes of a Fringe Watcher" that used to appear in *The Skeptical Inquirer*. Accompanying these columns are book reviews and pieces on a variety of topics. The book also includes my lengthy introduction to G. K. Chesterton's masterpiece of fantasy, *The Man Who Was Thursday*, and concludes with four articles that I contributed to Gordon Stein's *Encyclopedia of the Paranormal*.

The categories into which the chapters are divided are somewhat arbitrary. For example, my harsh look at the now happily forgotten Spiritualist religion based on a massive channeled book called *Oahspe* is here under the topic of religion. It could just as well be classified as moonshine.

Chapter 17 is included partly because it describes a bizarre exorcism, but mainly it is a hatchet job on Ernest Hemingway. Gertrude Stein, who disliked Hemingway, believed that F. Scott Fitzgerald was by far the better writer. She also thought F. Scott was more intelligent than Hemingway, and when he wasn't boozing, a habit which nearly killed him, he was a more admirable man. I find it gratifying that as Hemingway's reputation slowly declines, Fitzgerald's reputation is still rising.

I wish to thank Kendrick Frazier for the privilege of appearing regularly in the magazine he so ably edits, and my editor at W. W. Norton, Robert Weil, for accepting and expertly steering the book through production.

Martin Gardner

PART I

Science

1

Multiverses and Blackberries

There be nothing so absurd but that some philosopher
[or cosmologist?—M.G.] has said it.

—Cicero

The American philosopher Charles Sanders Peirce somewhere remarked that unfortunately universes are not as plentiful as blackberries. One of the most astonishing of recent trends in science is that many top physicists and cosmologists now defend the wild notion that not only are universes as common as blackberries, but even *more* common. Indeed, there may be an infinity of them!

It all began seriously with an approach to quantum mechanics (QM) called "The Many Worlds Interpretation" (MWI). In this view, widely defended by such eminent physicists as Murray Gell-Mann, Stephen Hawking, and Steven Weinberg, at every instant when a quantum measurement is made that has more than one possible outcome, the number specified by what is called the Schrödinger equation, the universe splits into two or more universes, each corresponding to a possible future. Everything that *can* happen at each juncture happens. Time is no longer linear. It is a rapidly branching tree. Obviously the number of separate universes increases at a prodigious rate.

If all these countless billions of parallel universes are taken as no more than abstract mathematical entities—worlds that could have formed but didn't—then the only "real" world is the one we are in. In this interpretation of the MWI the theory becomes little more than a new and whimsical language for talking about QM. It has the same mathematical formalism, makes the same predictions. This is how Hawking and many others who favor the MWI interpret it. They prefer it because they believe it is a language that simplifies QM talk, and also sidesteps many of its paradoxes.

There is, however, a more bizarre way to interpret the MWI. Those holding what I call the realist view actually believe that the endlessly sprouting new universes are "out there," in some sort of vast super-space-time, just as "real" as the universe we know! Of course at every instant a split occurs each of us becomes one or more close duplicates, each traveling a new universe. We have no awareness of this happening because the many universes are not causally connected. We simply travel along the endless branches of time's monstrous tree in a series of universes, never aware that billions upon billions of our replicas are springing into existence somewhere out there. "When you come to a fork in the road," Yogi Berra once said, "take it."

It is true that the MWI, in this realist form, avoids some of the paradoxes of QM. The so-called "measurement problem," for example, is no longer a problem because whenever a measurement occurs, there is no "collapse of the wave function" (or rotation of the state vector in a different terminology). All possible outcomes take place. Schrödinger's notorious cat is never in a mixed state of alive and dead. It lives in one universe, dies in another. But what a fantastic price is paid for these seeming simplicities! It is hard to imagine a more radical violation of Occam's razor, the law of parsimony which urges scientists to keep entities to a minimum.

The MWI was first put forth by Hugh Everett III in a Princeton doctoral thesis written for John Wheeler in 1956. It was soon taken up and elaborated by Bryce DeWitt. For several years John Wheeler defended his student's theory, but finally decided it was "on the wrong track," no more than a bizarre language for QM and one that carried "too much metaphysical baggage." However, recent polls show that about half of all QM experts now favor the theory, though it is seldom clear whether they think the other worlds are physically real or just abstractions such as numbers and triangles. Apparently both Everett and DeWitt took the realist approach. Roger Penrose is among many famous physicists who find the MWI appalling. The late Irish physicist John S. Bell called the MWI "grotesque" and just plain "silly." Most working physicists simply ignore the theory as nonsense.

In an article on "Quantum Mechanics and Reality" (in *Physics Today*, September 1970), DeWitt wrote with vast understatement about his first reaction to Everett's thesis: "I still recall vividly the shock I experienced on first encountering the multiworld concept. The idea of 10^{100+} slightly imperfect copies of oneself all constantly splitting into further copies, which ultimately become unrecognizable, is not easy to reconcile with common sense. This is schizophrenia with a vengeance!"

In the MWI, most of its defenders agree, there is no room for free will. The multiverse, the universe of all universes, develops strictly along determinist lines, always obeying the deterministically evolving Schrödinger equation. This equation is a monstrous wave function which never collapses unless it is observed and collapsed by an intelligence outside the multiverse, namely God.

In recent years David Deutsch, a quantum physicist at Oxford University, has become the top booster of the MWI in its realist form. He believes that quantum computers, using atoms or photons and operating in parallel with computers in nearby parallel worlds, can be trillions of times faster than today's computers. He is convinced that many famous QM paradoxes, such as the double slit experiment and a similar one involving two half-silvered mirrors, are best explained by assuming an interaction with twin particles in a parallel world almost identical with our own. For example, in the double slits experiment, when both slits are open, our particle goes through one slit while its twin from the other world goes through the other slit to produce the interference pattern on the screen.

Deutsch calls our particle the "tangible" one, and the particle coming from the other world a "shadow" particle. Of course in the adjacent universe our particle is the shadow of *their* tangible particle. Because communication between universes is impossible, it is hard to imagine why a particle would bother to jump from one universe to another just to produce interference.

Deutsch believes that the results of calculating simultaneously in parallel worlds can somehow be brought back here to coalesce. Critics argue that QM paradoxes, as well as quantum computers, are just as easily explained by conventional theory or by such rivals as the pilot wave theory of David Bohm. In any case, Deutsch's 1997 book *The Fabric of Reality: The Science of Parallel Universes—and Its Implications* is the most vigorous defense yet of a realistic MWI.

Deutsch is fully aware that the MWI forces him to accept the reality of endless copies of himself out there in the infinity of other worlds. "I may feel subjectively," he writes (p. 53), "that I am distinguished among the copies as the 'tangible' one, because I can directly perceive myself and not the others, but I must come to terms with the fact that all the others feel the same about themselves. Many of those Davids are at this moment writing these very words. Some are putting it better. Others have gone for a cup of tea." And he is puzzled by the fact that so few physicists are as enthralled as he about the MWI!

Theoretical and experimental work on quantum computers is now a complex, controversial, rapidly growing field with Deutsch as its pioneer and leading theoretician. You can keep up with this research by clicking on Oxford's Centre for Quantum Computation's Web site www.Qubic.org.

The MWI should not be confused with a more recent concept of a multiverse proposed by Andrei Linde, a Russian physicist now at Stanford University, as well as by a few other cosmologists such as England's Martin Rees. This multiverse is essentially a response to the anthropic argument that there must be a Creator because our universe has so many basic physical constants so finely tuned that, if any one deviated by a tiny fraction, stars and planets could not form—let alone life appear on a planet. The implication is that such fine tuning implies an intelligent tuner.

Linde's multiverse goes like this. Every now and then, whatever that means, a quantum fluctuation precipitates a Big Bang. A universe with its own space-time springs into existence with randomly selected values for its constants. In most of these universes those values will not permit the formation of stars and life. They simply drift aimlessly down their rivers of time. However, in a very small set of universes the constants will be just right to allow creatures like you and me to evolve. We are here not because of any overhead intelligent planning but simply because we happen by chance to be in one of the universes properly tuned to allow life to get started.

We come now to a third kind of multiverse, by far the wildest of the three. It has been set forth not by a scientist but by a peculiar philosopher, now at Princeton University, named David Lewis. In his best-known book, *The Plurality of Worlds* (Oxford, 1986), and other writings, Lewis seriously maintains that every logically possible universe—that is, one with no logical contradictions such as square circles—is somewhere out there. The notion of logical possible worlds, by the way, goes back to Leibniz's *Theodicy*. He speculated that God considered all logically possible worlds, then created the one He deemed best for His purposes.

Both the MWI and Lewis's possible worlds allow time travel into the past. You need never encounter the paradox of killing yourself, yet you are still alive, because as soon as you enter your past the universe splits into a new one in which you and your duplicate coexist.

Most of Lewis's worlds do not contain any replicas of you, but if they do they can be as weird as you please. You can't, of course, simultaneously have five fingers on each hand and seven on each hand because that would be logically contradictory. But you could have a hundred fingers, and a dozen arms, or seven heads. Any world you can think of without contradiction is real. Can pigs fly? Certainly. There is nothing contradictory about pigs with wings. In an infinity of possible worlds there are lands of Oz, Greek gods on Mount Olympus, *anything* you can imagine. Every novel is a possible world. Somewhere millions of Ahabs are chasing whales. Somewhere millions of Huckleberry Finns are floating down rivers. *Every* kind of universe exists if it is logically consistent.

David Lewis's mad multiverse was anticipated by hordes of science-fiction writers long before the MWI of QM came from Everett's brain. More recent examples include Larry Nivens's 1969 story "All the Myriad Ways" and Frederick Pohl's 1986 *The Coming of the Quantum Cats*. Jorge Luis Borges played with the theme in his story "The Garden of Forking Paths." There is a quotation from this tale at the front of *The Many Worlds Interpretation of Quantum Mechanics* (1973), a standard reference by DeWitt and Neill Graham. For other examples of multiverses in science fiction and fantasy see the entry on "Parallel Worlds" in *The Encyclopedia of Science Fiction* (1995) by John Clute and Peter Nichols.

Fredric Brown, in *What Mad Universe* (1950), described Lewis's multiverse this way:

> There are, then, *an infinite number of coexistent universes.*
>
> "They include this one and the one you came from. They are equally real, and equally true. But do you conceive what an infinity of universes means, Keith Winton?"
>
> "Well—yes and no."
>
> "It means that, out of infinity, *all conceivable universes exist.*
>
> "There is, for instance, a universe in which this exact scene is being repeated except that you—or the equivalent of you—are wearing brown shoes instead of black ones.
>
> "There are an infinite number of permutations of that variation, such as one in which you have a slight scratch on your left forefinger and one in which you have purple horns and—"
>
> "But are they all me?"
>
> Mekky said, "No, none of them is you—any more than the Keith Winton in this universe is you. I should not have used that pronoun. They are separate individual entities. As the Keith Winton here is; in this particular variation there is a wide physical difference—no resemblance, in fact."
>
> Keith said thoughtfully, "If there are infinite universes, then all possible combinations must exist. Then, somewhere, *everything must be true.*"
>
> "And there are an infinite number of universes, of course, in which we don't exist at all—that is, no creatures similar to us exist at all. In which the human race doesn't exist at all. There are an infinite number of universes, for instance, in which flowers are the predominant form of life—or in which no form of life has ever developed or will develop.
>
> "And infinite universes in which the states of existence are such that we would have no words or thoughts to describe them or to imagine them."

I have here looked at only the three most important versions of a multiverse. There are others, less well known, such as Penn State's Lee Smolin's universes which breed and evolve in a manner similar to Darwinian theory. For a good look at all the multiverses now being proposed, see British philosopher John Leslie's excellent book *Universes* (1989).

I find it hard to believe that so many academics take Lewis's possible worlds seriously. As poet Armand T. Ringer has put it in a clerihew:

David Lewis
Is a philosopher who is
Crazy enough to insist
That all logically possible worlds
actually exist.

Alex Oliver, reviewing Lewis's *Papers in Metaphysics and Epistemology*, in *The London Times Literary Supplement* (January 7, 2000), closes by calling Lewis "the leading metaphysician at the start of this century, head and beard above his contemporaries."

The stark truth is that there is not the slightest shred of reliable evidence that there is any universe other than the one we are in. No multiverse theory has so far provided a prediction that can be tested. In my layman's opinion they are all frivolous fantasies. As far as we can tell, universes are not as plentiful as even *two* blackberries. Surely the conjecture that there is just one universe and its Creator is infinitely simpler and easier to believe than that there are countless billions upon billions of worlds, constantly increasing in number and created by nobody. I can only marvel at the low state to which today's philosophy of science has fallen.

ADDENDUM

Martin Gardner Replies

A patient after a thorough physical examination: "How do I stand?"
Doctor: "That's what puzzles me."

—a HENNY YOUNGMAN joke

Bryce DeWitt, in *The Skeptical Inquirer* (March/April 2002) attacked my column in an article titled "Comments on Martin Gardner's

'Multiverses and Blackberries.' " He defended, as he has done in the past, his belief that the other worlds of the MWI are just as real as the universe we are in. "It was sad," he wrote, to see me "stumble" in my criticisms of the MWI. What follows here is my reply to DeWitt as it ran in *The Skeptical Inquirer* (May/June 2002).

David Lewis died in 2001, at age sixty. Sarah Boxer, in her long obit in the *New York Times* (October 20, 2001) quoted from my column. Let me add that aside from his eccentric views on possible worlds, Lewis made many notable contributions to many areas of philosophy.

Here is my reply:

I WAS PUZZLED by Bryce DeWitt's comments (*The Skeptical Inquirer*, March/April 2002) as well as honored that he would take seriously my column on "Multiverses and Blackberries" (September/October 2001). I know he read my column at least once because it so infuriated him, but did he read it again before he put forth his animadversions?

For example, he calls attention to the fact that three famous physicists, Steven Weinberg, Murray Gell-Mann, and Stephen Hawking, support the many worlds interpretation (MWI) of quantum mechanics. Did DeWitt forget that I wrote: "This view [is] widely defended by such eminent physicists as Murray Gell-Mann, Stephen Hawking, and Steven Weinberg"?

In what way do they defend it? The main point of my discussion of the MWI (only part of a column on multiverses in general) is that the MWI has two widely different interpretations, A and B.

A. The myriads of other worlds are not real in the *same way* as our world is real. They are abstract concepts in the MWI's language. This is the interpretation favored by Weinberg and Hawking. (I'm not sure of Gell-Mann's current belief.) Indeed, Hawking once said that the MWI is "trivially true." It is true in a way similar to the way the Pythagorean theorem is true even though you'll not find pure squares and triangles anywhere in the cosmos. "Is there any difference," DeWitt asks, between things "physically real" and "abstractions such as numbers and triangles"?

It is hard to believe a physicist could ask such a question. There is an enormous difference. Ten "exists" only as an abstraction. But ten pebbles and ten cows are "real" in an obviously different sense. There is a vast difference between the concept of a perfect triangle, and crude models cut from cardboard or drawn on paper.

B. The countless worlds of the MWI are every bit as physically "real" as the world we are now in. Although they are theoretical concepts, they are, like numbers, modeled by physically real entities such as billions of universes. This is the minority view held by such MWI experts as David Deutsch, from whose work I quoted, and by DeWitt himself.

DeWitt admits that these other worlds are not now observable. But this is also the case, he writes, with planets in distant galaxies. We believe they are there, even though we can't see them. Many physicists, he correctly adds (Mach for one) denied the reality of atoms. Now we "see" them with the aid of powerful instruments. Someday, DeWitt is convinced, the other worlds of the MWI may become observable! Clearly he believes that these worlds are "out there" and just as "real" as the universe we know.

DeWitt counters my objection that the MWI violates Occam's razor by arguing that it also uses the razor to cut down the number of *concepts* needed to talk about quantum mechanics. Did he forget that I said precisely this? I devoted a paragraph to how the MWI eliminates thorny paradoxes such as the paradox of Schrödinger's notorious cat. I explained how the MWI eliminates the "measurement problem" by doing away with Niels Bohr's "collapse of the wave function" by substituting a single wave function for the entire universe: one that evolves deterministically and never collapses. However, for such simplifications, I maintained, the MWI pays an incredible price.

One further point. Although DeWitt admits that the MWI is strictly deterministic, he thinks I "blundered" again by saying that this forbids free will. On this metaphysical question, I agree with William James and many contemporary thinkers that free will and determinism are incompatible. Interested readers can check the chapter on free will in my *Whys of a Philosophical Scrivener*.

2

A Skeptical Look at Karl Popper

Sir Karl Popper
Perpetrated a whopper
When he boasted to the world that he and he alone
Had toppled Rudolf Carnap from his Vienna Circle throne.

—a clerihew by ARMAND T. RINGER

Sir Karl Popper, who died in 1994, was widely regarded as England's greatest philosopher of science since Bertrand Russell, indeed a philosopher of worldwide eminence. Today his followers among philosophers of science are a diminishing minority, convinced that Popper's vast reputation is enormously inflated.

I agree. I believe that Popper's reputation was based mainly on his persistent but misguided efforts to restate commonsense views in a novel language that is rapidly becoming out of fashion.

Consider Popper's best known claim: that science does not proceed by "induction"—that is, by finding confirming instances of a conjecture—but rather by falsifying bold, risky conjectures. Confirmation, he argued, is slow and never certain. By contrast, a falsification can be sudden and definitive. Moreover, it lies at the heart of the scientific method.

A familiar example of falsification concerns the assertion that all crows are black. Every find of another black crow obviously confirms the theory, but there is always the possibility that a nonblack crow will turn up. If this happens, the conjecture is instantly discredited. The more often a conjecture passes efforts to falsify it, Popper maintained, the greater becomes its "corroboration," although corroboration is also uncertain and can never be quantified by a degree of probability. Popper's critics insist that "corroboration" is a form of induction, and Popper has simply sneaked induction in through a back door by giving it a new name. David Hume's famous question was "How can induction be justified?" It can't be, said Popper, because there *is* no such thing as induction!

There are many objections to this startling claim. One is that falsifications are much rarer in science than searches for confirming instances. Astronomers look for signs of water on Mars. They do not think they are making efforts to falsify the conjecture that Mars never had water.

Falsifications can be as fuzzy and elusive as confirmations. Einstein's first cosmological model was a universe as static and unchanging as Aristotle's. Unfortunately, the gravity of suns would make such a universe unstable. It would collapse. To prevent this, Einstein, out of thin air, proposed the bold conjecture that the universe, on its pre-atomic level, harbored a mysterious, undetected repulsive force he called the "cosmological constant." When it was discovered that the universe is expanding, Einstein considered his conjecture falsified. Indeed, he called it "the greatest blunder of my life." Today, his conjecture is back in favor as a way of explaining why the universe seems to be expanding faster than it should. Astronomers are not trying to falsify this "dark energy"; they are looking for confirmations.

Falsification may be based on faulty observation. A man who claims he saw a white crow could be mistaken or even lying. As long as observations of black crows continue, it can be taken in two ways: as confirmations of "all crows are black," or disconfirmations of "some crows are not black." Popper recognized, but dismissed as unimportant, that every falsification of a conjecture is simultaneously a confirmation of an opposite conjecture, and every confirming instance of a conjecture is a falsification of an opposite conjecture.

Consider the current hypothesis that there is a quantum field called the Higgs field, with its quantized particle. If a giant atom smasher some day, perhaps soon, detects a Higgs, it will confirm the conjecture that the field exists. At the same time it will falsify the opinion of some top physicists, Oxford's Roger Penrose for one, that there *is* no Higgs field.

To scientists and philosophers outside the Popperian fold, science operates mainly by induction (confirmation), but also and less often by disconfirmation (falsification). Its language is almost always one of induction. If Popper bet on a certain horse to win a race, and the horse won, you would not expect him to shout, "Great! My horse failed to lose!"

Astronomers are now finding compelling evidence that smaller and smaller planets orbit distant suns. Surely this is inductive evidence that there may be Earth-sized planets out there. Why bother to say, as each new and smaller planet is discovered, that it tends to falsify the conjecture that there are no small planets beyond our solar system? Why scratch your left ear with your right hand? Astronomers are looking for small planets. They are not trying to refute a theory any more than physicists are trying to refute the conjecture that there is no Higgs field. Scientists seldom attempt to falsify. They are inductivists who seek positive confirmations.

At the moment the wildest of all speculations in physics is superstring theory. It conjectures that all basic particles are different vibrations of extremely tiny loops of great tensile strength. No superstring has yet been observed, but the theory has great explanatory power. Gravity, for example, is implied as the simplest vibration of a superstring. Like prediction, explanation is an important aspect of induction. Relativity, for instance, not only made rafts of successful predictions but explained data previously unexplained. The same is true of quantum mechanics. In both fields researchers used classical induction procedures. Few physicists say they are looking for ways to falsify superstring theory. They are instead looking for confirmations.

Ernest Nagel, Columbia University's famous philosopher of science, in his *Teleology Revisited and Other Essays in the Philosophy and History of Science* (1979), summed it up this way: "[Popper's] conception of the role of falsification . . . is an oversimplification that is close to being a caricature of scientific procedures."

For Popper, what his chief rival, Rudolf Carnap, called a "degree of confirmation"—a logical relation between a conjecture and all relevant evidence—is a useless concept. Instead, as I said earlier, the more tests for falsification a theory passes, the more it gains in "corroboration." It's as if someone claimed that deduction doesn't exist, but of course statements can logically imply other statements. Let's invent a new term for deduction, such as "justified inference." It's not so much that Popper disagreed with Carnap and other inductivists as that he restated their views in a bizarre and cumbersome terminology.

To Popper's credit he was, like Russell, and almost all philosophers, scientists, and ordinary people, a thoroughgoing realist in the sense that he believed the universe, with all its intricate and beautiful mathematical structures, was "out there," independent of our feeble minds. In no way can the laws of science be likened to traffic regulations or fashions in dress that vary with time and place. Popper would have been as appalled as Russell by the crazy views of today's social constructivists and postmodernists, most of them French or American professors of literature who know almost nothing about science.

Scholars unacquainted with the history of philosophy often credit Popper for being the first to point out that science, unlike math and logic, is never absolutely certain. It is always corrigible, subject to perpetual modification. This notion of what the American philosopher Charles Peirce called the "fallibilism" of science goes back to ancient Greek skeptics, and is taken for granted by almost all later thinkers.

In *Quantum Theory and the Schism in Physics* (1982) Popper defends at length his "propensity theory" of probability. A perfect die, when tossed, has the propensity to show each face with equal probability. Basic particles, when measured, have a propensity to acquire, with specified probabilities, such properties as position, momentum, spin and so on. Here again Popper is introducing a new term which says nothing different from what can be better said in conventional terminology.

In my opinion Popper's most impressive work, certainly his best known, was his two-volume *The Open Society and Its Enemies* (1945). Its central theme, that open democratic societies are far superior to closed totalitarian regimes, especially Marxist ones, was hardly new, but Popper defends it with powerful arguments and awesome erudition. In later books he attacks what he calls "historicism," the belief that there are laws of historical change that enable one to predict humanity's future. The future is unpredictable, Popper argued, because we have free wills. Like William James, Popper was an indeterminist who saw history as a series of unforeseeable events. In later years he liked to distinguish between what he called three "worlds"—the external physical universe, the inner world of the mind, and the world of culture. Like Carnap and other members of the Vienna Circle, he had no use for God or an afterlife.

Karl Raimund Popper was born in Vienna in 1902 where he was also educated. His parents were Jewish, his father a wealthy attorney, his mother a pianist. For twenty years he was a professor of logic and scientific method at the London School of Economics. In 1965 he was knighted by the Crown.

I am convinced that Popper, a man of enormous egotism, was motivated by an intense jealousy of Carnap. It seems that every time Carnap expressed an opinion, Popper felt compelled to come forth with an opposing view, although it usually turned out to be the same as Carnap's but in different language. Carnap once said that the distance between him and Popper was not symmetrical. From Carnap to Popper it was small, but the other way around it appeared huge. Popper actually believed that the movement known as logical positivism, of which Carnap was leader, had expired because he, Popper, had single-handedly killed it!

I have not read Popper's first and only biography, *Karl Popper: The Formative Years (1902–1945)*, by Malachi Haim Hacohen (2000). Judging by the reviews it is an admirable work. David Papineau, a British philosopher, reviewed it for *The New York Times Book Review* (November 12, 2000). Here are his harsh words about Popper's character and work:

> By Hacohen's own account, Popper was a monster, a moral prig. He continually accused others of plagiarism, but rarely acknowledged his own intellectual debts. He expected others to make every sacrifice for him, but did little in return. In Hacohen's words, "He remained to the end a spoiled child who threw temper tantrums when he did not get his way." Hacohen is ready to excuse all this as the prerogative of genius. Those who think Popper a relatively minor figure are likely to take a different view.
>
> When Popper wrote "Logik der Forschung," he was barely thirty. Despite its flawed center, it was full of good ideas, from perhaps the most brilliant of the bright young philosophers associated with the Vienna Circle. But where the others continued to learn, develop and

in time exert a lasting influence on the philosophical tradition, Popper knew better. He refused to revise his falsificationism, and so condemned himself to a lifetime in the service of a bad idea.

Popper's great and tireless efforts to expunge the word *induction* from scientific and philosophical discourse has utterly failed. Except for a small but noisy group of British Popperians, induction is just too firmly embedded in the way philosophers of science and even ordinary people talk and think. Confirming instances underlie our beliefs that the Sun will rise tomorrow, that dropped objects will fall, that water will freeze and boil, and a million other events. It is hard to think of another philosophical battle so decisively lost.

Readers interested in exploring Popper's eccentric views will find, in addition to his books and papers, most helpful the two-volume *Philosophy of Karl Popper* (1970), in the Library of Living Philosophers, edited by Paul Arthur Schilpp. The book contains essays by others, along with Popper's replies and an autobiography. For vigorous criticism of Popper, see David Stove's *Popper and After: Four Modern Irrationalists* (the other three are Imre Lakatos, Thomas Kuhn, and Paul Feyerabend), and Stove's chapter on Popper in his posthumous *Against the Idols of the Age* (1999), edited by Roger Kimball. See also Carnap's reply to Popper in *The Philosophy of Rudolf Carnap* (1963), another volume in The Library of Living Philosophers. Of many books by Popperians, one of the best is *Critical Rationalism* (1994), a skillful defense of Popper by his top acolyte.

ADDENDUM

Because there is so little space for commenting on the many letters about this column this ran in *The Skeptical Inquirer* (November/December 2001), I will limit my remarks to Harry White's main point. He chides me for not mentioning Popper's claim that a theory has no cognitive content unless it can be falsified, and that this provides a useful basis for demarcating good from bad science.

I omitted this aspect of Popper's philosophy of science because the idea was not new. It was advanced by earlier thinkers, notably William Whewell and Charles Peirce. Popper's favorite example of a theory that cannot be falsified was Freudian analysis. This is surely wrong. Adolf Grünbaum, in *The Foundations of Psychoanalysis* (1984) shows clearly that psychoanalysis not only can be falsified, but that in fact it already has been.

3

Rudolf Carnap, Philosopher of Science

When I was a freshman at the University of Chicago in 1932, I intended to transfer after two years to the California Institute of Technology to become a physicist. For better or worse, I got sidetracked into philosophy for my bachelor's degree. After one year of graduate work on a scholarship, I decided not to continue for the master's degree but to become a writer instead. I had a job in the university's press-relations office when I enlisted in the Navy.

Back in Chicago after four years of service as a yeoman, I used the G.I. bill in the fall of 1946 to take a seminar with Rudolf Carnap. Titled "Concepts, Theories and Methods in the Physical Sciences," it was the most exciting class I ever attended. It led me into a lifelong interest in the philosophy of science.

After each session I typed the notes I had taken and put them in a looseleaf binder, to which I added an index. Looking over the introductory page in the binder, I find a record reminding me that, during the hour prior to Carnap's class, the same room was used for a course about the "Great Books." (The university was then in its notorious Robert

Hutchins–Mortimer Adler phase, which stressed the classics of the Western world.) This often left the blackboard covered with diagrams explaining some aspect of Plato's or Aristotle's metaphysics. Carnap never looked at these diagrams while he erased them. I have likened this sweep of Carnap's arm across the blackboard to his erasure of stale and meaningless metaphysics. One day, before Carnap arrived, a student did the erasing, explaining that it was "so as not to worry Carnap."

Carnap's carefully constructed sentences were delivered slowly, in a low, rich, pleasing voice, giving the impression that he was struggling to simplify ideas too complex for us to fully understand. A favorite phrase, after describing a difficult task, was "Now how could we do that?" Other often-used phrases included "and we know this for the following reasons," "blurs the distinction," "lacks cognitive content," and "prescientific thinking."

Carnap opened each session with a summary of what he had said at the previous meeting, followed by a period of questions. The philosopher most often cited was Carnap's good friend Hans Reichenbach, with Carl Hempel running second.

I was surprised that Carnap seldom mentioned Bertrand Russell, although I knew he owed Russell a great debt. Later I attended a seminar given by Russell on the Chicago campus. Carnap was in the audience and asking questions. Much of their give and take was beyond me, though I recall Russell saying at one point, "But realism is not a dirty word." They had been arguing over whether it is desirable for a philosopher to assume the reality of an external world as something ontologically certain, or whether realism is no more than the most convenient, indeed indispensable, language for science. Carnap liked to call it the "thing language." Russell soon turned this into a question of whether Carnap's wife was truly "out there" or should be regarded merely as a useful construction within Carnap's experience.

I have written elsewhere about what occurred the next day. I was in the university's post office, talking to philosopher Charles Hartshorne, when Carnap strode in. To my eternal embarrassment, Hartshorne said to Carnap, "Mr. Gardner has been telling me that during Russell's seminar yesterday he tried to persuade you that your wife existed, but you wouldn't admit it."

Carnap glowered at me and said, "But that wasn't the point at all."

Many years later, when Carnap repeated his seminar at the University of California at Los Angeles, I wrote to propose a book. The plan was for someone—who would turn out to be Carnap's wife, Ina—to attend the seminar and tape-record each session. She would then type out everything he said, including questions and answers, and send me the pages after each typing. I would edit the material into a coherent volume, working questions and replies into the text as best I could. By then I had started my writing career, and Carnap was familiar with some of my work. He liked the proposal, and the result was a book at first titled *The Philosophical Foundations of Physics* (Basic Books, 1966), and later, as a Dover paperback (1995), *An Introduction to the Philosophy of Science.* Every idea in the book is Carnap's. Only the phrasing and arrangements are mine. The time I spent working on this book, as copy went back and forth between me and Carnap for corrections and clarifications, was one of the happiest periods of my life.

Although I never knew Carnap personally and never met his wife, I have vivid and fond memories of his class. He was a teacher who always did his best to make a question, no matter how stupid, seem significant, and to extract from it a meaningful comment. His lectures were extemporaneous, though based on notes he carried on file cards.

It was during this course that Carnap shocked us all by revealing that his friend and former associate Moritz Schlick had just been murdered by a psychotic student. There were, however, moments of comedy as well. I remember one confusing interchange with a woman mathematics teacher before it was discovered that she was using the word *pear*, the fruit: Carnap had taken it to be *pair*—or maybe it was the other way around.

Basic Books published our book in 1966 under Carnap's preferred title. After Carnap's death in 1970, Basic reissued the book in paper covers, and the original subtitle became the new title, which has been retained in the Dover edition. Many corrections for the Basic edition were generously supplied by Carnap's friend Carl G. Hempel. I had asked Hempel to write a foreword, but he declined because he considered it inappropriate to write a foreword to a book by a person so much more eminent than he.

The book received good reviews, and was adopted in the classroom by Wesley Salmon and a few other noted philosophers of science here and abroad. Translations were published in Germany, France, Italy, Japan, and Argentina. It should be emphasized that this is the only book by Carnap on a level sufficiently nontechnical to be understood by readers with no expertise in mathematics, physics, or logic. . . .

It was a great privilege to have attended Carnap's seminar and to have been given the honor of editing his book. Although Carnap's reputation is not as high now as it was then, I have no doubt that it will steadily rise again. More and more younger philosophers of science will surely discover the greatness of his contributions and his influence, and how right he was in his notable quarrels (in my opinion largely verbal quibbles) with Karl Popper and Willard Van Orman Quine.

4

Some Thoughts About Induction

Ask a scientist what he conceives the scientific method to be, and he will adopt an expression that is at once solemn and shifty-eyed: solemn, because he feels he ought to declare an opinion; shifty-eyed, because he is wondering how to conceal the fact that he has no opinion to declare.

—Sir Peter Brian Medawar,
in *Induction and Intuition in Scientific Thought* (1969)

"Induction" has several meanings. Here I shall confine it to the process by which science, on the basis of intuitive hunches followed by empirical evidence, arrives at laws and theories about how the universe operates. Unlike mathematical and logical deduction, it is never absolutely certain, but when the evidence is overwhelming, a law or theory is often called a fact. For example, the constancy of the speed of light regardless of the observer's motion was originally a cornerstone of Einstein's special theory of relativity. It has now been verified to such a high degree that physicists refer to it as a fact. What was once Darwin's theory of evolution is now regarded as a fact by everybody except religious fundamentalists.

Induction operates in two ways. It either advances a conjecture by what are called confirming instances, or it falsifies a conjecture by contrary or disconfirming evidence. A common example is the hypothesis that all crows are black. Each time a new crow is observed and found to be black the conjecture is increasingly confirmed. But if a crow is found to be not black the conjecture is falsified.

You might suppose that the term "confirming instance" can be precisely defined. Although in most cases it is easily recognized there are many cases where the term becomes extremely fuzzy. The most notorious example, known as Hempel's paradox, was discovered by the German philosopher Carl Gustav Hempel. It goes like this: The statement "All crows are black" is logically equivalent to "No not-black object is a crow." This seems to allow an ornithologist to do research in his living room. He observes, say, a blue vase and notes it is not a crow. This clearly is a confirming instance of "No not-black object is a crow." But if it confirms that statement, does it not also confirm the logically equivalent statement not all crows are black?

The answer is clearly yes if we limit our universe to a small set of black and not-black objects. For example, suppose the "universe" consists of just two shoe boxes on a shelf so high that one can reach into each box but not see into it. A "scientist" is told only that box A contains six cubes, and box B contains six red objects.

The scientist reaches into box A and removes one cube at a time. Each time he gets a black cube it confirms the assertion that all six cubes are black. After he has seen all six has the conjecture been verified? No, because box B may contain a red cube. So, the scientist takes out red objects, one by one. Each time it is seen not to be a cube it surely confirms the statement "No not-black object is a cube," and therefore also confirms the statement that all six cubes are black.

However, in the world of real crows the number of not-black objects is close to infinity. Nevertheless, Hempel argued, the observation of a blue vase or a red apple does indeed confirm "All crows are black," but only to an infinitesimal degree. Other philosophers of science disagree. They maintain that seeing a blue vase adds nothing to the evidence that all crows are black. The debate is still the topic of academic papers. If you care to look into this far-from-settled controversy you will find a lengthy bibliography on pages 191–94 of my *Hexaflexagons and Other Mathematical Diversions* (1988).

Hempel's paradox is not the only example of how fuzzy the term "confirming instance" can be. I will give only one more instance. (For others, see Wesley Salmon's admirable article, "Confirmation," in *Scientific American*, May 1973.) Consider the conjecture "A person can grow to fifteen feet tall." Now imagine that in some remote jungle a man is discovered who is fourteen-and-a-half feet tall. From one point of view he does not confirm the conjecture because he fails by a half a foot. On the other hand he can be seen as a confirming instance because if a man can grow fourteen-and-a-half feet it becomes much more likely that he could grow another six inches.

How can induction be justified? David Hume considered this question and concluded that induction has no logical justification. No matter how many times the sun rises, there is no *logical* reason why it might not fail to rise tomorrow. The same applies to any given law of nature. For all we know, it could suddenly alter. Today's philosophers of science all agree with Hume. But can induction be justified in some nonlogical way? Hans Reichenbach defended it by saying that the only conceivable way we can learn anything about nature is by making inductions from available evidence. This is certainly true, but it's not much of a justification.

David Hume, and later John Stuart Mill, gave what has become the most popular justification. Induction works, they said, because nature is orderly; that is, it is based on mathematical patterns that are assumed to hold throughout the universe and not to change with time. Once this "uniformity of nature," as Mill called it, is granted, it becomes plausible that observing part of a pattern will justify assuming that the pattern will generalize into a law. As Hume and many later philosophers pointed out, this involves circular reasoning. We justify induction by assuming that nature is uniform, but the only way we decide it is uniform is by induction. However, as Max Black and others argued, in this case circular reasoning is not vicious.

Bertrand Russell, in his early years, thought it possible that some day a logical basis for induction could be found. In his later years he abandoned this hope. His last major work, *Human Knowledge, Its Scope and Limits* (1948), reached the same conclusion as Hume and Mill. Here is how he summarized the book in his introduction:

> Inference from a group of events to other events can only be justified if the world has certain characteristics which are not logically necessary. So far as deductive logic can show, any collection of events might be the whole universe; if, then, I am ever to be able to infer

events, I must accept principles of inference which lie outside deductive logic. All inference from events to events demands some kind of interconnection between different occurrences. Such interconnection is traditionally asserted in the principle of causality or natural law. It is implied, as we shall find, in whatever limited validity may be assigned to induction by simple enumeration. But the traditional ways of formulating the kind of interconnection that must be postulated are in many ways defective, some being too stringent and some not sufficiently so. To discover the minimum principles required to justify scientific inferences is one of the main purposes of this book.

Richard Feynman, on a *Nova* television interview in 1983, likened the scientist to a man who knows nothing about chess, but is allowed once every day to observe for one minute a small part of a chessboard while a game is in progress. It may take him many weeks to realize that bishops move only along diagonals and rooks move only along orthogonals. Months later he may think he knows how all the pieces move and capture, but then come surprises. He sees a pawn captured *en passant*, or a king and rook castle. A big surprise is seeing that a player has two queens on the board. It could take the scientist years to learn that a pawn on the eighth row can become a queen, rook, bishop, or knight, or exactly what checkmates and stalemates are.

The main point of the chess analogy is that induction works only on universes that are uniformly patterned. The chess universe is very tiny, restricted to an eight-by-eight checkered square and a small number of pieces and rules. The actual universe is playing a game on a far vaster scale. For all we know the space-time "board" may be infinite in both space and time, and there could be an infinity of rules. But without positing the uniformity of nature, induction would be hopeless. $E=mc^2$ both on Earth and in distant galaxies. All electrons, fortunately, are identical. They may be made of superstrings. If so, what are superstrings made of? Maybe the universe is infinite in both directions.

A number of induction games have been invented that model the process of guessing a general rule by observing part of a pattern. One is the card game called Eleusis, created by Robert Abbott and described in my *Penrose Tiles to Trapdoor Ciphers* (1989). Another induction game

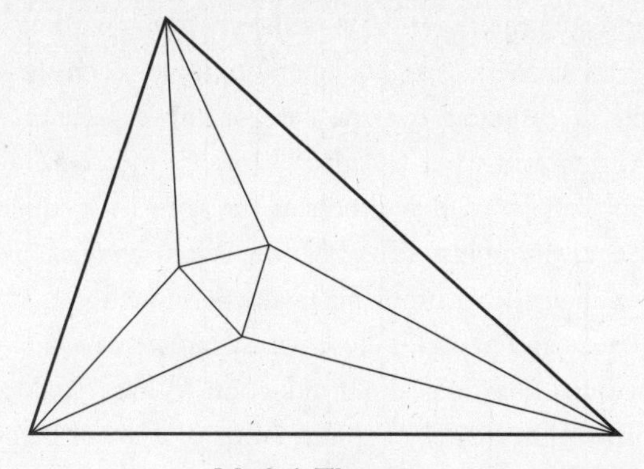

Morley's Theorem

called Patterns, played on a board, was invented by Sidney Sackson and described in my book *Mathematical Circus* (1979). An intriguing variant of Eleusis, by mathematician Martin Kruskal, can be found in his 1962 booklet, *Delphi—A Game of Induction*. Efforts are being made here and there to program computers to make inductions when fed statements of evidence, but so far they have not discovered any new laws, though they have rediscovered some old ones.

In his later years Rudolf Carnap wrote a big book titled *Logical Foundations of Probability* in which he defined "degree of confirmation" as a logical relation between a conjecture and all relevant evidence. He believed that in the future it would be possible to construct an inductive logic based on a formal language in which a hypothesis and its evidence could be precisely stated. The degree of confirmation could then be expressed by a probability between one and zero, a value that would alter as the evidence altered. Most other philosophers of science, notably Sir Karl Popper, have disagreed. They think it will never be possible to quantify a degree of confirmation. Of course scientists often express a degree of confirmation by giving their subjective estimate. Rival efforts to construct inductive logics have been made by other philosophers, but the efforts are primitive.

It is worth noting that mathematicians often discover a theorem by experimenting with numbers or diagrams in a way that closely resembles the way scientists induce a physical law. For example, Frank Morley, father of the writer Christopher Morley, was surprised one day to find that if he trisected the angles of an arbitrary triangle, three intersection points were the corners of an equilateral triangle. Each time he experimented with a different triangle he got the same beautiful result. Eventually he proved what is called "Morley's theorem." The point is that every time a new triangle produced the same little equilateral triangle it was a confirming instance of Morley's conjecture.

There are many other problems involving induction that are covered in endless books and papers. I close with what seems to me the deepest of all unanswered questions about induction. What role does simplicity play in adding to the degree of confirmation of a law or theory? Given a choice between two conjectures for which the known evidence is equal, it seems that the simpler of the two is more likely to be fruitful. Is this always the case? Alfred North Whitehead once advised, "Seek simplicity but distrust it." And can a conjecture's simplicity be quantified? Is there a way to measure simplicity? So far no one has come close to answering these questions.

The problem is more than academic. Suppose a laboratory is considering two competing theories for which the evidence is equal. To test each theory would involve equipment costing a million dollars per theory. But how the simplicity could be determined is not clear, and no one knows why the simpler theory is more likely to be true.

5

Can Time Go Backward?

. . . time, dark time, secret time, forever flowing like a river. . . .

—THOMAS WOLFE,
The Web and the Rock

Time has been described by many metaphors, but none is older or more persistent than the image of time as a river. You cannot step twice in the same river, said Heraclitus, the Greek philosopher who stressed the temporal impermanence of all things, because new waters forever flow around you. You cannot even step into it once, added his pupil Cratylus, because while you step both you and the river are changing into something different. As Ogden Nash put it in his poem "Time Marches On,"

While ladies draw their stockings on,
The ladies they were are up and gone.

In James Joyce's *Finnegans Wake* the great symbol of time is the river Liffey flowing through Dublin, its "hither-and-thithering waters" reaching the sea in the final lines, then returning to "riverrun," the book's first word, to begin again the endless cycle of change.

It is a powerful symbol, but also a confusing one. It is not time that flows but the world. "In what units is the rate of time's flow to be measured?" asked the Australian philosopher J. J. C. Smart. "Seconds per ———?" To say "time moves" is like saying "length extends." As Austin Dobson observed in his poem "The Paradox of Time,"

Time goes, you say? Ah no!
Alas, time stays, we go.

Moreover, whereas a fish can swim upriver against the current, we are powerless to move into the past. The changing world seems more like the magic green carpet that carried Ozma across the Deadly Desert (the void of nothingness?), unrolling only at the front, coiling up only at the back, while she journeyed from Oz to Ev, walking always in one direction on the carpet's tiny green region of "now." Why does the magic carpet never roll backward? What is the physical basis for time's strange, undeviating asymmetry?

THERE HAS BEEN as little agreement among physicists on this matter as there has been among philosophers. Now, as the result of recent experiments, the confusion is greater than ever. Before 1964 all the fundamental laws of physics, including relativity and quantum laws, were "time-reversible." That is to say, one could substitute $-t$ for t in any basic law and the law would remain as applicable to the world as before; regardless of the sign in front of t the law described something that could occur in nature. Yet there are many events that are possible in theory but that never or almost never actually take place. It was toward those events that physicists turned their attention in the hope of finding an ultimate physical basis for distinguishing the front from the back of "time's arrow."

A star's radiation, for example, travels outward in all directions. The reverse is never observed: radiation coming from all directions and converging on a star with backward-running nuclear reactions that make it an energy sink instead of an energy source. There is nothing in the basic laws to make such a situation impossible in principle; there is only the difficulty of imagining how it could get started. One would have to assume that God or the gods, in some higher continuum, started the waves at the rim of the universe. The emergence of particles from a disintegrating radioactive nucleus and the production of ripples when a stone is dropped into a quiet lake are similar instances of one-way events. They never occur in reverse because of the enormous improbability that "boundary conditions"—conditions at the "rim" of things—would be such as to produce the required kind of converging energy. The reverse of beta decay, for instance, would require that an electron, a proton, and an antineutrino be shot from the "rim" with such deadly accuracy of aim that all three particles would strike the same nucleus and create a neutron.

The steady expansion of the entire cosmos is another example. Here again there is no reason why this could not, in principle, go the other way. If the directions of all the receding galaxies were reversed, the red shift would become a blue shift, and the total picture would violate no known physical laws. All these expanding and radiative processes, although always one-way as far as our experience goes, fail to provide a fundamental distinction between the two ends of time's arrow.

IT HAS BEEN suggested by many philosophers, and even by some physicists, that it is only in human consciousness, in the one-way operation of our minds, that a basis for time's arrow can be found. Their arguments have not been convincing. After all, the earth had a long history before any life existed on it, and there is every reason to believe that earthly events were just as unidirectional along the time axis then as they are now. Most physicists came finally to the conclusion that all natural events are time-reversible in principle (this became known technically as "time invariance") except for events involving the statistical behavior of large numbers of interacting objects.

Consider what happens when a cue ball breaks a triangle of fifteen balls on a pool table. The balls scatter hither and thither and the eight ball, say, drops into a side pocket. Suppose immediately after this event the motions of all the entities involved are reversed in direction while keeping the same velocities. At the spot where the eight ball came to rest the molecules that carried off the heat and shock of impact would all converge on the same spot to create a small explosion that would start the ball back up the incline. Along the way the molecules that carried off the heat of friction would move toward the ball and boost it along its upward path. The other balls would be set in motion in a similar fashion. The eight ball would be propelled out of the side pocket and the balls would move around the table until they finally converged to form a triangle. There would be no sound of impact because all the molecules that had been involved in the shock waves produced by the initial break of the triangle would be converging on the balls and combining with their momentum in such a way that the impact would freeze the triangle and shoot the cue ball back toward the tip of the cue. A motion picture of any individual molecule in this event would show absolutely nothing unusual. No basic mechanical law would seem to be violated. But when the billions of "hither-and-thithering" molecules involved in the total picture are considered, the probability that they would all move in the way required for the time-reversed event is so low that no one can conceive of its happening.

Because gravity is a one-way force, always attracting and never repelling, it might be supposed that the motions of bodies under the influence of gravity could not be time-reversed without violating basic laws. Such is not the case. Reverse the directions of the planets and they would swing around the sun in the same orbits. What about the collisions of objects drawn together by gravity—the fall of a meteorite, for example? Surely *this* event is not time-reversible. But it is! When a large meteorite strikes the earth, there is an explosion. Billions of molecules scatter hither and thither. Reverse the directions of all those molecules and their impact at one spot would provide just the right amount of energy to send the meteorite back into orbit. No basic laws would be violated, only statistical laws.

IT WAS HERE, in the laws of probability, that most nineteenth-century physicists found an ultimate basis for time's arrow. Probability explains such irreversible processes as the mixing of coffee and cream, the breaking of a window by a stone, and all the other familiar one-way-only events in which large numbers of molecules are involved. It explains the second law of thermodynamics, which says that heat always moves from hotter to cooler regions, increasing the entropy (a measure of a certain

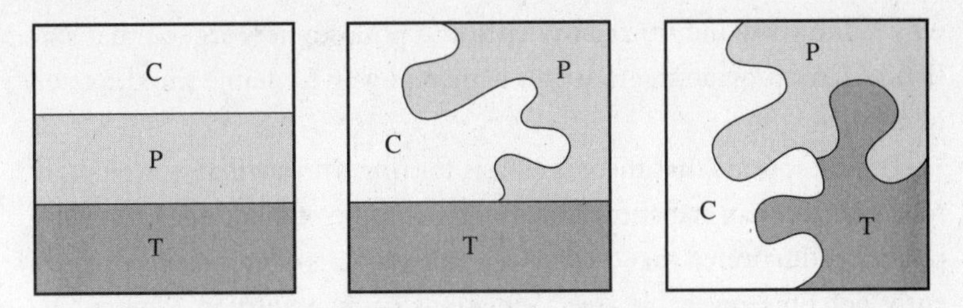

Three symmetries, charge (C), *parity* (P), *and time* (T), *are likened to pieces that fit into a pattern. Before 1957 they were all assumed to be symmetrical; any experiment (the pattern) involving the three could be duplicated with any one piece, any two or all three reversed* (left). *Then experiments were found that violate* P*-symmetry, suggesting that if overall* (CPT) *symmetry holds, some piece other than* P *must also be asymmetrical.* C *was found to be such a piece; an experiment remains the same if* C *and* P *are reversed together* (middle). *In 1964 experiments that violate this* CP*-symmetry were reported. It follows that* T *must be asymmetrical in these cases, since a pattern violating* CP*-symmetry can be duplicated only by reversing all three pieces simultaneously* (right).

kind of disorder) of the system. It explains why shuffling randomizes a deck of ordered cards.

"Without any mystic appeal to consciousness," declared Sir Arthur Eddington (in a lecture in which he first introduced the phrase "time's arrow"), "it is possible to find a direction of time. . . . Let us draw an arrow arbitrarily. If as we follow the arrow we find more and more of the random element in the state of the world, then the arrow is pointing towards the future; if the random element decreases the arrow points towards the past. That is the only distinction known to physics."

Eddington knew, of course, that there are radiative processes, such as beta decay and the light from suns, that never go the other way, but he did not consider them sufficiently fundamental to provide a basis for time's direction. Given the initial and boundary conditions necessary for starting the reverse of a radiative process, the reverse event is certain to take place. Begin with a deck of disordered cards, however, and the probability is never high that a random shuffle will separate them into spades, hearts, clubs and diamonds. Events involving shuffling processes seem to be irreversible in a stronger sense than radiative events. That is why Eddington and other physicists and philosophers argued that statistical laws provide the most fundamental way to define the direction of time.

It now appears that there is a basis for time's arrow that is even more fundamental than statistical laws. In 1964 a group of Princeton University physicists discovered that certain weak interactions of particles are apparently not time-reversible [see "Violations of Symmetry in Physics," by Eugene P. Wigner; *Scientific American*, December 1965]. One says "apparently" because the evidence is both indirect and controversial. Although it is possible to run certain particle interactions backward to make a direct test of time symmetry, such experiments have not as yet shown any violations of time-reversibility. The Princeton tests were of an indirect kind. They imply, if certain premises are granted, that time symmetry is violated.

The most important premise is known as the *CPT* theorem. *C* stands for electric charge (plus or minus), *P* for parity (left or right mirror images) and *T* for time (forward or backward). Until a decade ago physicists believed each of these three basic symmetries held throughout nature. If you reversed the charges on the particles in a stone, so that plus charges became minus and minus charges became plus, you would still have a stone. To be sure, the stone would be made of antimatter, but there is no reason why antimatter cannot exist. An antistone on the earth would instantly explode (matter and antimatter annihilate each other when they come in contact), but physicists could imagine a galaxy of antimatter that would behave exactly like our own galaxy; indeed, it could be in all respects exactly like our own except for its *C* (charge) reversal.

The same universal symmetry was believed to hold with respect to *P* (parity). If you reversed the parity of a stone or a galaxy—that is, mirror-reflected its entire structure down to the last wave and particle—the result would be a perfectly normal stone or galaxy. Then in 1957 C. N. Yang and T. D. Lee received the Nobel prize in physics for theoretical work that led to the discovery that parity is *not* conserved [see "The Overthrow of Parity," by Philip Morrison; *Scientific American*, April

1957]. There are events on the particle level, involving weak interactions, that cannot occur in mirror-reflected form.

IT WAS AN unexpected and disturbing blow, but physicists quickly regained their balance. Experimental evidence was found that if these asymmetrical, parity-violating events were reflected in a special kind of imaginary mirror called the *CP* mirror, symmetry was restored. If in addition to ordinary mirror reflection there is also a charge reversal, the result is something nature can "do." Perhaps there are galaxies of antimatter that are also mirror-reflected matter. In such galaxies, physicists speculated, scientists could duplicate every particle experiment that can be performed here. If we were in communication with scientists in such a *CP*-reversed galaxy, there would be no way to discover whether they were in a world like ours or in one that was *CP*-reflected. (Of course, if we went there and our spaceship exploded on arrival, we would know we had entered a region of antimatter.)

No sooner had physicists relaxed a bit with this newly restored symmetry than the Princeton physicists found some weak interactions in which *CP* symmetry appears to be violated. In different words, they found some events that, when *CP*-reversed, are (in addition to their *C* and *P* differences) not at all duplicates of each other. It is at this point that time indirectly enters the picture, because the only remaining "magic mirror" by which symmetry can be restored is the combined *CPT* mirror in which all three symmetries—charge, parity, and time—are reversed. This *CPT* mirror is not just something physicists want to preserve because they love symmetry. It is built into the foundations of relativity theory in such a way that, if it turned out not to be true, relativity theory would be in serious trouble. There are therefore strong grounds for believing the *CPT* theorem holds. *On the assumption that it does*, a violation of *CP* symmetry would imply that time symmetry is also violated [see illustration on page 33]. There are a few ways to preserve the *CP* mirror without combining it with *T*, but none has met with any success. The best way is to suppose there is a "fifth force" (in addition to the four known forces: gravity, the weak-interaction force, electromagnetism, and the nuclear force) that is causing the newly discovered anomalies. Experiments have cast strong doubt on the fifth-force hypothesis, however.

In 1967 Paolo Franzini and his wife, working with the alternating-gradient synchrotron at the Brookhaven National Laboratory, found even stronger evidence of *CP* violations—this time in events involving electromagnetic reactions. The Franzini work was controverted, however, by a group of physicists at the European Organization for Nuclear Research (CERN) in Geneva, who announced their results in September. At the moment the cause of this discrepancy in results is not clear.

Although the evidence is still indirect and in part controversial, many physicists are now convinced that there are events at the particle level that go in only one time direction. If this holds throughout the universe, there is now a way to tell, while communicating with scientists in a distant galaxy, whether they are in a world of matter or of antimatter. We simply ask them to perform one of the *CP*-violating experiments. If their description of such a test coincides exactly with our own description of the same test when done here, we shall not explode when we visit them. It may well be that the universe contains no galaxies of antimatter. But physicists like to balance things, and if there is as much antimatter as there is matter in the universe, there may be regions of the cosmos in which all three symmetries are reversed. Events in our world that are lopsided with respect to *CPT* would all go the other way in a *CPT*-reversed galaxy. Its matter would be mirror-reflected, reversed in charge and moving backward in time.

WHAT DOES IT mean to say that events in a galaxy are moving backward in time? At this point no one really knows. The new experiments indicate that there is a preferred time direction for certain particle interactions. Does this arrow have any connection with other time arrows such as those that are defined by radiative processes, entropy laws, and the psychological time of living organisms? Do all these arrows have to point the same way or can they vary independently in their directions?

Before the recent discoveries of the violation of *T* invariance the most popular way to give an operational meaning to "backward time" was by imagining a world in which shuffling processes went backward, from disorder to order. Ludwig Boltzmann, the nineteenth-century Austrian physicist who was one of the founders of statistical thermodynamics, realized that after the molecules of a gas in a closed, isolated container have reached a state of thermal equilibrium—that is, are moving in complete disorder with maximum entropy—there will always be little pockets forming here and there where entropy is momentarily decreasing. These would be balanced by other regions where entropy is increasing; the overall entropy remains relatively stable, with only minor up-and-down fluctuations.

Boltzmann imagined a cosmos of vast size, perhaps infinite in space and time, the overall entropy of which is at a maximum but which contains pockets where for the moment entropy is decreasing. (A "pocket" could include billions of galaxies and the "moment" could be billions of years.) Perhaps our fly-speck portion of the infinite sea of space-time is one in which such a fluctuation has occurred. At some time in the past, perhaps at the time of the "big bang," entropy happened to decrease; now it is increasing. In the eternal and infinite flux a bit of order happened to put in its appearance; now that order is disappearing again, and so our arrow of time runs in the familiar direction of increasing entropy. Are there other regions of space-time, Boltzmann asked, in which the arrow of entropy points the other way? If so, would it be correct to say that time in such a region was moving backward, or should one simply say that entropy was decreasing as the region continued to move forward in time?

It seems evident today that one cannot speak of backward time without meaning considerably more than just a reversal of the entropy arrow. One has to include all the other one-way processes with which we are familiar, such as the radiative processes and the newly discovered *CP*-violating interactions. In a world that was completely time-reversed *all* these processes would go the other way. Now, however, we must guard against an amusing verbal trap. If we imagine a cosmos running backward while we stand off somewhere in space to observe the scene, then we must be observing the cosmos moving backward in a direction opposite to our own psychological time, which still runs forward. What does it mean to say that the *entire* cosmos, including all possible observers, is running backward?

In the first book of Plato's *Statesman* a stranger explains to Socrates his theory that the world goes through vast oscillating cycles of time. At the end of each cycle time stops, reverses and then goes the other way. This is how the stranger describes one of the backward cycles:

> The life of all animals first came to a standstill, and the mortal nature ceased to be or look older, and was then reversed and grew young and delicate; the white locks of the aged darkened again, and the cheeks of the bearded man became smooth, and recovered their former bloom; the bodies of youths in their prime grew softer and smaller, continually by day and night returning and becoming assimilated to the nature of a newly born child in mind as well as body; in the succeeding stage they wasted away and wholly disappeared.

Plato's stranger is obviously caught in the trap. If things come to a standstill in time and "then" reverse, what does the word *then* mean? It has meaning only if we assume a more fundamental kind of time that continues to move forward, altogether independent of how things in the universe move. Relative to this meta-time—the time of the hypothetical observer who has slipped unnoticed into the picture—the cosmos is indeed running backward. But if there *is* no meta-time—no observer who can stand outside the entire cosmos and watch it reverse—it is hard to understand what sense can be given to the statement that the cosmos "stops" and "then" starts moving backward.

There is less difficulty—indeed, no logical difficulty at all—in imagining two portions of the universe, say two galaxies, in which time goes one way in one galaxy and the opposite way in the other. The philosopher Hans Reichenbach, in his book *The Direction of Time*, suggests that this could be the case, and that intelligent beings in each galaxy would regard their own time as "forward" and time in the other galaxy as "backward." The two galaxies would be like two mirror images: each would seem reversed to inhabitants of the other. From this point of view time is a relational concept like up and down, left and right or big and small. It would be just as meaningless to say that the *entire* cosmos reversed its time direction as it would be to say that it turned upside down or suddenly became its own mirror image. It would be meaningless because there is no absolute or fixed time arrow outside the cosmos by which such a reversal could be measured. It is only when *part* of the cosmos is time-reversed in relation to another part that such a reversal acquires meaning.

NOW, HOWEVER, WE come up against a significant difference between mirror reflection and time reversal. It is easy to observe a reversed world—one has only to look into a mirror. But how could an observer in one galaxy "see" another galaxy that was time-reversed? Light, instead of radiating from the other galaxy, would seem to be going toward it. Each galaxy would be totally invisible to the other. Moreover, the memories of observers in the two galaxies would be operating in opposite directions. If you somehow succeeded in communicating something to someone in a time-reversed world, he would promptly forget it because the event would instantly become part of his future rather than of his past. "It's a poor sort of memory that only works backward," said Lewis Carroll's White Queen in one looking-glass, time-reversed (*PT*-reversed!) scene. Unfortunately, outside of Carroll's dream world, memory works only one way. Norbert Wiener, speculating along such lines in his book *Cybernetics*, concluded that no communication would be possible between intelligent beings in regions with opposite time directions.

A British physicist, F. Russell Stannard, pursues similar lines of thought in an article on "Symmetry of the Time Axis" (*Nature*, August 13, 1966) and goes even further than Wiener. He concludes (and not all physicists agree with him) that no interactions of any kind would be possible between particles of matter in two worlds whose time axes pointed in opposite directions. If the universe maintains an overall symmetry with respect to time, matter of opposite time directions would "decouple" and the two worlds would become invisible to each other. The "other" world "would consist of galaxies absorbing their light rather than emitting it, living organisms growing younger, neutrons being created in triple collisions between protons, electrons and antineutrinos, and thereafter being absorbed in nuclei, etc. It would be a universe that was in a state of contraction, and its entropy would be decreasing, and yet the faustian observers ["faustian" is Stannard's term for the "other"

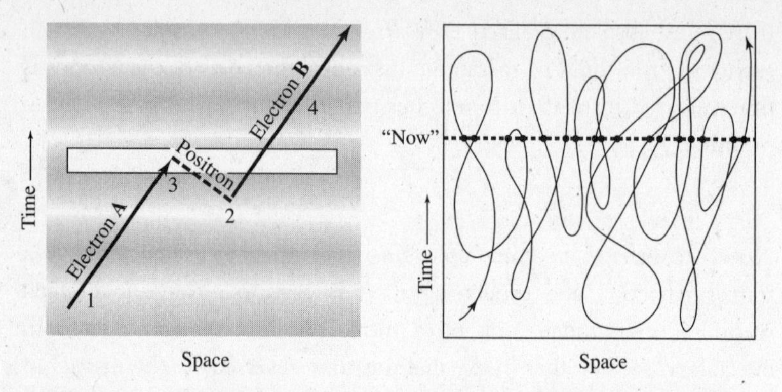

Feynman graph, shown at the left in a simplified form devised by Banesh Hoffman of Queens College, shows how an antiparticle can be considered a particle moving backward in time. The graph is viewed through a horizontal slot in a sheet of cardboard (gray) *that is moved slowly up across the graph. Looking through the slot, one sees events as they appear in our forward-looking "now." Electron* A *moves to the right* (1), *an electron-positron pair is created* (2), *the positron and electron* A *are mutually annihilated* (3) *and electron* B *continues on to the right* (4). *From a timeless point of view (without the slotted cardboard), however, it can be seen that there is only one particle: an electron that goes forward in time, backward and then forward again. Richard P. Feynman's approach stemmed from a whimsical suggestion by John A. Wheeler of Princeton University: A single particle, tracing a "world line" through space and time* (right), *could create all the world's electrons* (black dots) *and positrons* (gray dots).

region] would not be aware of anything strange in their environment. Being constructed of faustian matter, their subjective experience of time is reversed, so they would be equally convinced that it was they who grew older and their entropy that increased."

Instead of one universe with oscillating time directions, as in the vision of Plato's stranger, Stannard's vision bifurcates the cosmos into side-by-side regions, each unrolling its magic carpet of history simultaneously (whatever "simultaneously" can mean!) but in opposite directions. Of course, there is no reason why the cosmos has to be symmetrical in an overall way just to satisfy the physicist's aesthetic sense of balance. The universe may well be permanently lopsided in regard to all three aspects—charge, parity, and time—even if there is no theoretical reason why all three could not go the other way. If a painting does not have to be symmetrical to be beautiful, why should the universe?

IS IT POSSIBLE to imagine a single individual living backward in a time-forward world? Plato's younger contemporary, the Greek historian Theopompus of Chios, wrote about a certain fruit that, when eaten, would start a person growing younger. This, of course, is not quite the same thing as a complete reversal of the person's time. There have been several science-fiction stories about individuals who grew backward in this way, including one amusing tale, "The Curious Case of Benjamin Button," by (of all people) F. Scott Fitzgerald. (It first appeared in *Colliers* in 1922 and is most accessible at the moment in *Pause to Wonder*, an anthology edited by Marjorie Fischer and Rolfe Humphries.) Benjamin is born in 1860, a seventy-year-old man with white hair and a long beard. He grows backward at a normal rate, enters kindergarten at sixty-five, goes through school and marries at fifty. Thirty years later, at the age of twenty, he decides to enter Harvard, and he is graduated in 1914 when he is sixteen. (I am giving his biological ages.) The Army promotes him to brigadier general because as a biologically older man he had been a lieutenant colonel during the Spanish-American War, but when he shows up at the Army base as a small boy he is packed off for home. He grows younger until he cannot walk or talk. "Then it was all dark," reads Fitzgerald's last sentence, "and his white crib and the dim faces that moved above him, and the warm sweet aroma of the milk, faded out altogether from his mind."

Aside from his backward growth, Mr. Button lives normally in forward-moving time. A better description of a situation in which the time arrows of a person and the world point in opposite directions is found in Carroll's novel *Sylvie and Bruno Concluded*. The German Professor hands the narrator an Outlandish Watch with a "reversal peg" that causes the outside world to run backward for four hours. There is an amusing description of a time-reversed dinner at which "an empty fork is raised to the lips: there it receives a neatly cut piece of mutton, and swiftly conveys it to the plate, where it instantly attaches itself to the mutton already there." The scene is not consistent, however. The order of the dinner-table remarks is backward, but the words occur in a forward time direction.

If we try to imagine an individual whose entire bodily and mental processes are reversed, we run into the worst kind of difficulties. For one thing, he could not pass through his previous life experiences backward, because those experiences are bound up with his external world, and since that world is still moving forward none of his past experiences can be duplicated. Would we see him go into a mad death dance, like an automaton whose motor had been reversed? Would he, from his point of view, find himself still thinking forward in a world that seemed to be going backward? If so, he would be unable to see or hear anything in that world, because all sound and light waves would be moving toward their points of origin.

We seem to encounter nothing but nonsense when we try to apply different time arrows to an individual and the world. Is it perhaps possible, on the microlevel of quantum theory, to speak sensibly about part of the universe moving the wrong way in time? It is. In 1948 Richard P. Feynman, who shared the 1965 Nobel prize in physics, developed a mathematical approach to quantum theory in which an antiparticle is regarded as a particle moving backward in time for a fraction of a microsecond. When there is pair-creation of an electron and its antiparticle the positron (a positively charged electron), the positron is extremely short-lived. It immediately collides with another electron, both are annihilated, and off goes a gamma ray. Three separate particles—one positron and two electrons—seem to be involved. In Feynman's theory there is only *one* particle, the electron [see illustration on page 40]. What we observe as a positron is simply the electron moving momentarily back in time. Because our time, in which we observe the event, runs uniformly forward, we see the time-reversed electron as a positron. We think the positron vanishes when it hits another electron, but the vanishing is at just the spot in time where the electron entered the past. The electron executes a tiny zigzag dance in space-time, hopping into the past just long enough for us to see its path in a bubble chamber and interpret it as a path made by a positron moving forward in time.

Feynman got his basic idea when he was a graduate student at

Princeton, from a telephone conversation with his physics professor, John A. Wheeler. In his Nobel-prize acceptance speech Feynman told the story this way:

"Feynman," said Wheeler, "I know why all electrons have the same charge and the same mass."

"Why?" asked Feynman.

"Because," said Wheeler, "they are all the *same* electron!"

Wheeler went on to explain on the telephone the stupendous vision that had come to him. In relativity theory physicists use what are called Minkowski graphs for showing the movements of objects through space-time. The path of an object on such a graph is called its "world line." Wheeler imagined one electron, weaving back and forth in space-time, tracing out a single world line. The world line would form an incredible knot, like a monstrous ball of tangled twine with billions on billions of crossings, the "string" filling the entire cosmos in one blinding, timeless instant. If we take a cross section through cosmic space-time, cutting at right angles to the time axis, we get a picture of three-space at one instant of time. This three-dimensional cross section moves forward along the time axis, and it is on this moving section of "now" that the events of the world execute their dance. On this cross section the world line of the electron, the incredible knot, would be broken up into billions on billions of dancing points, each corresponding to a spot where the electron knot was cut. If the cross section cuts the world line at a spot where the particle is moving forward in time, the spot is an electron. If it cuts the world line at a spot where the particle is moving backward in time, the spot is a positron. All the electrons and positrons in the cosmos are, in Wheeler's fantastic vision, cross sections of the knotted path of this single particle. Since they are all sections of the same world line, naturally they will all have identical masses and strengths of charge. Their positive and negative charges are no more than indications of the time direction in which the particle at that instant was weaving its way through space-time.

There is an enormous catch to all this. The number of electrons and positrons in the universe would have to be equal. You can see this by drawing on a sheet of paper a two-dimensional analogue of Wheeler's vision. Simply trace a single line over the page to make a tangled knot [see illustration on page 40]. Draw a straight line through it. The straight line represents a one-dimensional cross-section at one instant in time through a two-space world (one space axis and one time axis). At points where the knot crosses the straight line, moving up in the direction of time's arrow, it produces an electron. Where it crosses the line going the opposite way it produces a positron. It is easy to see that the number of electrons and positrons must be equal or have at most a difference of one. That is why, when Wheeler had described his vision, Feynman immediately said:

"But, Professor, there aren't as many positrons as electrons."

"Well," countered Wheeler, "maybe they are hidden in the protons or something."

Wheeler was not proposing a serious theory, but the suggestion that a positron could be interpreted as an electron moving temporarily backward in time caught Feynman's fancy, and he found that the interpretation could be handled mathematically in a way that was entirely consistent with logic and all the laws of quantum theory. It became a cornerstone in his famous "space-time view" of quantum mechanics, which he completed eight years later and for which he shared his Nobel prize. The theory is equivalent to traditional views, but the zigzag dance of Feynman's particles provided a new way of handling certain calculations and greatly simplifying them. Does this mean that the positron is "really" an electron moving backward in time? No, that is only one physical interpretation of the "Feynman graphs"; other interpretations, just as valid, do not speak of time reversals. With the new experiments suggesting a mysterious interlocking of charge, parity, and time direction, however, the zigzag dance of Feynman's electron, as it traces its world line through space-time, no longer seems as bizarre a physical interpretation as it once did.

AT THE MOMENT no one can predict what will finally come of the new evidence that a time arrow may be built into some of the most elementary particle interactions. Physicists are taking more interest than ever before in what philosophers have said about time, thinking harder than ever before about what it means to say time has a "direction" and what connection, if any, this all has with human consciousness and will. Is history like a vast "riverrun" that can be seen by God or the gods from source to mouth, or from an infinite past to an infinite future, in one timeless and eternal glance? Is freedom of will no more than an illusion as the current of existence propels us into a future that in some unknown sense already exists? To vary the metaphor, is history a prerecorded motion picture, projected on the four-dimensional screen of our spacetime for the amusement or edification of some unimaginable Audience?

Or is the future, as William James and others have so passionately argued, open and undetermined, not existing in *any* sense until it actually happens? Does the future bring genuine novelty—surprises that even the gods are unable to anticipate? Such questions go far beyond the reach of physics and probe aspects of existence that we are as little capable of comprehending as the fish in the river Liffey are of comprehending the city of Dublin.

References

Chown, Marcus. "Can Time Run Backwards?" *New Scientist*, February 5, 2000.

Davies, Paul. "Backwards in Time," Chapter 10 in *About Time* (Simon & Schuster, 1995).

MacBeath, Murray. "Communication and Time Reversal." *Synthese*, Vol. 56 (1983).

Zee, A. "Time Reversal." *Discover*, October 1992. Reprinted in Zee's *Mysteries of Life and the Universe* (1993).

Three science-fiction novels deal with reverse time: Brian Aldiss, *Cryptozoic* (1967); Philip K. Dick, *Counter-Clock World* (1967); and Martin Amis, *Time's Arrow* (1991). *Time's Arrow* was reviewed by John Updike in *The New Yorker*, May 25, 1992.

PART II

Mathematics

6

Against the Odds

This short story appeared in *The College Mathematics Journal* (Vol. 22, January 2001).

Luther Washington was a friendly, shy, intelligent boy, the oldest of five children who lived with their parents in Butterfield, Kansas. He was one of sixteen African-Americans who attended Butterfield Central High. His father owned a small grocery store in the town's black district. His mother cooked dinners for one of the town's bankers.

For some reason, which his parents never understood, Luther was fascinated by numbers. One day, when he was ten, he surprised his father by saying: "Dad, I've discovered an easy way to tell if a big number can be divided by four or eight and not have anything left over."

The family was having breakfast. "Tell us about it," said Mr. Washington.

"Well," said Luther, "if the last two digits"—his grade-school teacher had taught him to distinguish digits from numbers—"can be divided by four with nothing left over, the big number also can be divided by four. Otherwise, it can't. And if the big number's last three digits are a product of eight, so is the big number."

"Interesting," said Mrs. Washington. "But I can't see how that could be of any use to anybody."

"I don't care whether it's any use or not. I just think it's neat."

"I wish you wouldn't waste so much time on such things," said Mr. Washington. "All the math you'll ever need in life is knowing how to add, subtract, multiply, and divide, and how to make change."

"Maybe so," said Luther. "But there's something wonderful and mysterious about numbers."

In high school Luther was quick to learn algebra and geometry even though his teacher, Miss Perkins, seemed bored by what she taught. The fact is, she really *was* bored. Her major, at a Kansas teachers' college, was American history, but when one of the school's two math teachers quit to take a better-paying job as a used-car salesman, Miss Perkins was asked to take over his classes. She reluctantly agreed. By studying the textbook carefully each night she managed fairly well to stay ahead of her students.

Miss Perkins was a devout Baptist. She felt no hostility toward blacks in general. After all, they were God's children and she believed that all persons are equally precious in the eyes of the Lord. On the other hand, she was convinced that the intelligence of blacks was considerably below that of whites. It had to be something in their genes.

Among the forty-two students in her geometry class, five were black—three girls and two boys. They would sit silently at their desks, their dark eyes fixed on hers. They rarely asked a question. When they did it was usually, in her opinion, a foolish one. If she asked who had solved a problem, she almost never called on a black student whose hand was raised.

One morning, after Miss Perkins had demonstrated on the blackboard how Euclid proved that the three angles of a triangle add to a straight angle, Luther was the only student to wave a hand. His classmates looked as if they were falling asleep.

"Yes, Luther, do you have a question?"

"No. But I thought of a simpler way to prove the theorem."

Miss Perkins raised her eyebrow and peered at Luther over her bifocals. "Would you like to come to the blackboard and explain it to the class?"

Luther stood up and walked to the blackboard. He erased the diagram Miss Perkins had chalked. After drawing a large triangle with irregular sides, he took a pencil from his shirt pocket and pressed it against the blackboard along the inside of the triangle's base.

"As you can see," he said, "the pencil points to the right." Luther slid the pencil along the triangle's base until its point touched the triangle's right corner. Keeping the point fixed at the corner, he swung the pencil clockwise until it was alongside the triangle's right side. He then moved the pencil upward along the side until its point touched the triangle's top corner. Again he rotated the pencil clockwise until it paralleled the triangle's left side. He slid the pencil down to the triangle's lower left corner, and rotated the pencil once more. It was now back where it started.

"See," Luther said as he turned his head to face the class, "the pencil now points left. Obviously it has rotated a hundred and eighty degrees. This proves that the three angles add to a straight angle."

The students were all now wide awake. About a dozen, including the three black girls, clapped. The others sat silent, waiting for Miss Perkins's reaction.

Miss Perkins looked a bit flustered. She knew how to demonstrate the theorem by folding over the corners of paper triangles, but Luther's method caught her by surprise. "I'm not sure that's really a proof," she said finally.

"Well, it is," said Luther. "The pencil acts like a vector. The best thing about it is you can apply it to any polygon, with any number of sides. It proves the inside angles must add to a multiple of a straight angle. And the same thing is true about any polygon's exterior angles."

Miss Perkins wasn't sure she grasped what Luther was saying. "What are vectors and exterior angles?" she thought. Fortunately, the bell rang and the class was over before she could make any other comments. "That boy is going to be a troublemaker," she said to herself. "He's not nearly as smart as he thinks he is."

A few days later, during a study period, Miss Perkins walked by Luther's desk. He had been doodling on a sheet of paper. Scattered over the sheet were numerous ticktacktoe patterns, their cells filled with Xs and Os.

"You're supposed to be working on your assignment."

"I've already finished it, Miss Perkins. I'm trying to figure out whether the first player can always win at ticktacktoe if both players do their best. Maybe the game's a draw. I do know the second player can never win, no matter how big the field is or how many marks he has to get in a straight line."

Miss Perkins snatched up the sheet and crumpled it into a ball. "When you're in my class," she said sternly, "I expect you to work on mathematics and nothing else. If you've finished your assignment, come to my desk and I'll give you some more problems."

"Yes, Miss Perkins."

To Miss Perkins's chagrin, Luther had near perfect scores on every test she gave. But she was so annoyed by what she considered his "uppity" way of talking to her that she gave him a B grade for the semester.

In the middle of each school year Butterfield Central High held contests to decide who among its seniors excelled in each of the school's subjects. Winners were given a medal and made members of an honorary society called the B Club.

Luther entered the math contest. Miss Perkins was annoyed when she learned that he was the first student in the B Club's history to solve every problem correctly. The test had been designed and the papers graded by the school's principal.

Members of the B Club met each Friday, after school. Miss Perkins was the club's sponsor. When Luther showed up at the next meeting, she told him he had been appointed chairman of the club's peanut-sacking committee.

"Who appointed me?" Luther asked.

"I did."

"What am I supposed to do?"

"You'll be in charge of a group of club members. They meet after school every Friday to make sure *all* the peanuts are put in paper sacks to sell at Saturday basketball games."

"How long will that take?"

"About two hours."

Luther groaned. Time after school was precious. He had found in the town's library a book titled *Mathematics and the Imagination*. It was an exciting book, and he was eager to get home to finish it.

"I don't want to sack peanuts," Luther said. "Am I allowed to resign from the B Club?"

Miss Perkins was so furious when Luther stalked out of the room that next Monday she sent him to the principal's office with a note saying he had refused to sack peanuts, and that he had been very unpleasant in the way he had spoken to her.

Now it so happened that the principal knew a great deal more about mathematics than Miss Perkins. Math had been his undergraduate major at the University of Michigan before he went on to do graduate work in education.

The principal had been impressed by the unusual ways Luther solved the problems on the test he had prepared. After discussing several mathematical topics with Luther, he realized he was talking to a young man with deep insights into mathematical structures.

"Are you planning on college?"

"No, sir. My parents can't afford it. Dad wants me to work in his grocery store. When he's too old to run the store, he expects me to take it over."

"How are your other grades?"

"Not so good. I barely passed Latin with a D. And I got a C in English. I didn't understand any of Shakespeare's plays. I was so bored by *Ivanhoe* that I couldn't finish it."

"Hmm." The principal stared at the ceiling for a while before he asked: "Have you applied for any scholarships?"

Luther shook his head. "My folks don't want me to go to college. They say it would be a big waste of time. Dad says if I had a college degree it would only make me overqualified for any kind of work I could get."

"What do you make of Miss Perkins as a teacher?"

Luther hesitated for a moment. "She's okay, I guess. I don't think she likes mathematics, though. I know she doesn't like me."

Luther told the principal about the time she had grabbed the sheet on which he was analyzing ticktacktoe, and how she told him he had to work only on math during her study periods.

The principal laughed, then sighed and looked solemn. "She means well. But she should have known that ticktacktoe is based on combinatorics and graph theory. Did you finally solve the game?"

"I did when I got home. I drew a complete game tree. It's not hard to do if you take advantage of symmetries like rotations and reflections. The game's a draw if both sides make their best moves."

"That's right. You may be interested to know that Frank Harary, a well known graph theorist, generalized the game in an interesting way. Have you head about polyominoes?"

Luther shook his head.

"They're shapes made by attaching unit squares along their edges. In standard ticktacktoe the goal is to be the first to form what is called a straight tromino. That's three little squares in a row. Harary varied the goal by making it any kind of polyomino on a field of any size. For example, suppose the goal is to be the first to form the only other tromino, three squares joined to make a right angle. It's called the bent tromino. If a bent tromino is the goal, who wins? Or is it also a draw?"

"I wish I'd thought of that," said Luther.

"Harary also analyzed reverse games for low-order polyominoes. Who wins if the first player forced to form a specified polyomino is the loser?"

"I can see how that could get very complicated," said Luther. "I did think of analyzing the reverse game for standard ticktacktoe. It's a draw, too. The first player can't win but he can force a draw by going first in the center."

The principal was interested. "Can you show me how he does that?"

For the next ten minutes Luther and the principal played reverse games of ticktacktoe.

"The chairman of the math department at Stanford University is an old friend," said the principal, as he pushed the sheets of paper to one side of his desk. "We were frat brothers at Michigan. If you don't mind, I'd like to send him your contest answers. Would you object if Stanford offered you a full scholarship that included room and board?"

"No."

"Would your parents object?"

"I don't know."

The chairman of Stanford's math department needed only a glance at Luther's unconventional ways of solving the test problems to know that he wanted Luther in his department. Even before he had a long phone conversation with the principal of Butterfield Central High, the chairman had started the process for Luther to be given a full freshman scholarship.

To Luther's vast surprise and joy, his mother and father agreed to let him accept the scholarship. He almost cried when his father embraced him and said how proud he was!

Luther's scholarship was renewed each year, and on through graduate work until he got his doctorate. Three years later he was made a full professor at Stanford. By then he had published a dozen technical papers, and was much in demand as a lecturer at math conventions.

Fast-forward ten years to 1970. Luther was awarded the Fields Medal, the equivalent in mathematics of a Nobel prize. It was for his brilliant solution of a long-standing conjecture in Diophantine analysis.

A news release about the award went to the *Butterfield Times*. The paper ran the story with Luther's picture on the front page.

Miss Perkins had retired from teaching after she married Mr. Jones, the school's basketball coach. During breakfast, her husband handed her the newspaper's front section.

"It says here that Professor Washington grew up in Butterfield and attended Central High. He must have been in one of your classes."

Miss Perkins adjusted her spectacles and studied Dr. Washington's picture for a full minute. "I don't recognize him," she said, frowning. "There were lots of black boys in my classes. They all looked so much alike. I wonder if he really deserved that prize. You know how affirmative action works these days. Would he have been given that award if he'd been white?"

ADDENDUM

Frank Harary's generalization of ticktacktoe to polyomino patterns was the topic of my *Scientific American* column, April 1979. The column is reprinted as Chapter 17 in *Fractal Music, Hypercards, and More* (W. H. Freeman, 1992).

The proof that the second player of ticktacktoe cannot win if each side plays rationally is beautifully simple. Assume the second player has a winning strategy. The first player makes an irrelevant move on any cell.

He thus becomes the second player and can stall the second player's assumed winning strategy! If this strategy requires the first player's irrelevant first move, he simply makes another irrelevant move. (Extra Xs on the field can only help, never hinder.)

The original assumption is thus contradicted and shown false. This clever proof was first formulated by John Nash for the game of Hex, which was independently invented in Denmark by Piet Hein. Because Hex cannot be drawn, the proof shows that the first player can always win, though it does not show how.

7

Fun with Möbius Bands

Some people think it takes years of experience or advanced calculating skills to explore complex mathematical ideas, but it doesn't. You and your children can do it if you have some paper, a little tape, and a pair of scissors. With those materials you can do experiments in topology, an intriguing branch of modern mathematics that deals with shapes and structures.

Topologists analyze the properties of an object that don't change even if you bend, stretch or distort it. For example, they might look at how many faces the object has, or how many holes there are in it. Like other mathematicians, topologists perform calculations, but their primary goal is really to understand how shapes work.

To get started, cut several thin strips of paper. Mark some with lines down the middle, and others with lines a third of the way in from each edge [see illustration on page 58]. Pick up one strip and take a look at it. It has two faces, front and back. Now tape the ends together to form a band. The band still has two separate faces. One way a "face" is defined in topology is as something separated from another face by an edge. If you trace your finger around one face of your band, you cannot get to the other face unless you cross an edge. Try it. The number of faces and edges the band has are some of its topological properties: Those numbers would not change no matter how much you stretched, folded, or distorted the band, as long as you didn't cut it or tape it again.

Do you think you can make a band with only one face and one edge? It sounds impossible, but the surprising answer is yes. Pick up another strip of paper and give it a half-twist before taping the ends together. The band looks like it has two faces, but don't be fooled. Run your finger along the face and you will see that you can trace the entire surface without crossing an edge. You've made a one-sided object! (The band also has only one edge.)

How is this possible? The secret lies in the half-twist, which connected the original strip's two faces—front to back and back to front—to make one face. (When you made the ordinary band, you connected each face to itself, which still gave the band two faces.)

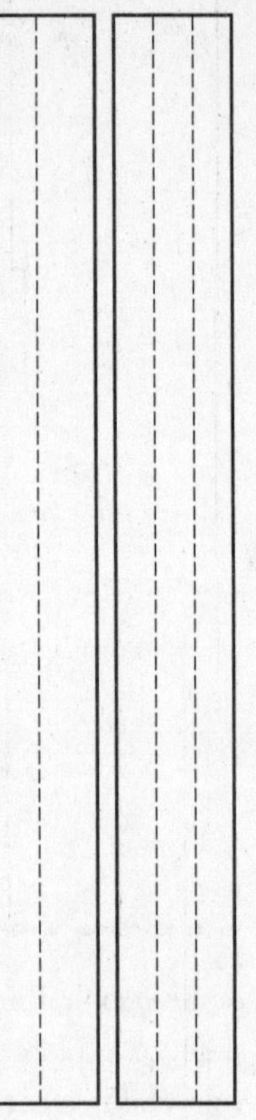

Cut out and mark strips like these to make your Möbius bands

Topologists call this kind of band a Möbius strip, after Augustus Ferdinand Möbius, a German mathematician and astronomer who was among the first to study one-sided surfaces.

The Möbius strip is one of the great curiosities of topology. In addition to its single face, it has a number of other unusual properties that arise when you try to cut it up. If you cut it down the middle, you would expect it to split into two strips. But it doesn't: instead, it turns into a single large strip, twice as long as the original. Try it—make your band with a strip that has a line down the middle to guide your cut. As an anonymous limerick has it:

A mathematician confided
That a Möbius strip is one-sided.

You'll get quite a laugh
If you cut it in half,
For it stays in one piece when divided.

Try cutting up another Möbius band, this time in thirds. Make the band with a strip that has two lines on it, and cut along the lines. After cutting around the band once, you'll notice that your scissors don't meet up with the beginning of the cut and you have to go around again to complete the job. It's hard to guess in advance what the outcome will be: You get one large band and one smaller band, linked together.

Next, try giving a strip of paper two half-twists before you join the ends. Is it still a Möbius surface? No. Run your finger along it and you'll soon discover that you have to cross an edge to get from one face to the other: the band has two faces. If you cut it down the middle, you'll get two linked bands of equal size. Try it.

If you make a band with three half-twists, you'll be back to an object with one face and one edge. Bisecting it produces a single band tied in a knot. If you experiment with greater numbers of twists, you'll find that when the number of half-twists is odd, the band will be one-sided, and bisecting it will always make a single band. If the number of half-twists is even, the band will be two-sided, and bisecting will always result in two linked bands. These are topological properties of the twisted bands. Can you find a pattern that links the number of twists in the band you started with to the number of twists in the bands that result when you cut it up?

Now make another topological toy. Cut two strips of the same size and put one on top of the other. Give the pair a half-twist, then join the ends to make a doubled Möbius band. Is this one strip or two? It sure looks like two. Indeed, you can "prove" that there are two separate Möbius strips by inserting a finger or a playing card "between" the strips and running it all the way around, back to where you started. If you open up this doubled band, however, you'll discover it is a single, two-sided strip; can you figure out why? (Trying to put the double band back into its original form is not easy!) Bisect the new band and you get two bands that are hopelessly intertwined.

For a final experiment, cut out a large cross from a sheet of paper.

Draw a guide line down the center of one arm, and label that arm "A." Draw two lines down the other arm, each a third of the way in from the edge. Label that arm "B." Join the ends of arm A to make a band with no twists. Give one end of arm B a half-twist and join the ends to make a Möbius surface.

Now trisect arm B, and then bisect arm A. (Be sure to do the steps in that order.) It's a safe bet that no one in your family will be able to predict the shape of the finished piece. Try it, and be astounded!

The Möbius strip has been featured in many science-fiction stories, and it even has some applications in technology. For example, many conveyor belts are made as Möbius strips, so that both "sides" wear evenly as they pass over the rollers. (A plain band would wear out on only one side.) Can you come up with your own variations of shapes and cuts?

8

Is Mathematics "Out There"?

With very few exceptions mathematicians have always believed, and still believe, that mathematical truths have a strange kind of abstract reality that is discovered, not created. In recent years a tiny minority of maverick mathematicians have joined the postmodern ranks of the social constructivists who see both math and science as cultural artifacts, unrelated to any sort of timeless truth domain. Reuben Hersh, a distinguished mathematician, has long defended this antirealist view, notably in his 1997 book *What Is Mathematics, Really?*

If all Hersh means is that mathematics is part of human culture, then of course he is right, but the statement is vacuous. Everything humans say and do is part of culture. Hersh obviously means something less trivial. He is convinced that mathematicians do not discover timeless theorems—theorems true in all possible worlds. Rather they create ever-changing, uncertain conjectures in much the same way that others create art, music, religion, wars, and traffic regulations. In an article in *Eureka* (March 1988), he discusses several "myths" which he claims have been mistakenly defended by great mathematicians.

No one can quarrel with Hersh's first myth, that Euclid put plane geometry on a firm formal foundation. All mathematicians today agree that he did not.

Myth 2: "Mathematical truth or knowledge is the same for everyone. It does not depend on who in particular discovers it; in fact, it is true whether or not anyone discovers it."

What a strange contention! In no culture on earth, or anywhere else, is the Pythagorean theorem not certain within the formal system of Euclidian geometry.

Consider a heap of n pebbles. The number is prime only if, when you take away pebbles in increments of k each (k not 1 or n), there always will be one or more pebbles left over. Are we allowed to say that n is either prime or composite before anyone makes such a test? If Hersh agrees, does this not give his game away?

Primality is a timeless property of certain integers, as independent of humanity as pebbles and stars. Humans are not needed to test a pile of pebbles for primality. It can be done by monkeys or even mindless machines. To reply to this by saying that humans are necessary to have the *concept* of a prime and to call numbers prime is to say something utterly trivial.

Myth 3: "If the little green men (and women?) from Quasar .X9 sent us their math textbooks, we would find again $A = \pi r^2$."

If the aliens haven't advanced to plane geometry, their textbooks of course would not contain this theorem, but if they knew about circles and areas how could they not discover that a circle's area is πr^2? Given the axioms of plane geometry, the theorem holds in all possible worlds.

Myth 4: "Mathematics possesses a method called 'proof' . . . by which one attains absolute certainty of conclusions, given the truth of the premises."

Can Hersh be serious when he calls this a myth? In mathematics, unlike in science, proof is the essence. Given the symbols, and the formation and transformation rules of a formal system, all theorems are tautologies. They are, as Kant was the first to say, analytic, not synthetic. Even in the center of the sun, Bertrand Russell once wrote, two plus two equals four.

I once put it this way. If two dinosaurs joined two other dinosaurs in a clearing, there would be four there even though no humans were around to observe it, and the beasts were too stupid to know it. Mathematical structure was deeply embedded in the universe long before sentient life evolved. Indeed, the structure was there a microsecond after the big bang, and even before the bang because there had to be quantum fields to fluctuate and explode. Is Hersh willing to say that galaxies had a spiral structure before creatures were around to use the word "spiral"?

No mathematician, Roger Penrose has observed, probing deeper into the intricate structure of the Mandelbrot set, can imagine he is not exploring a pattern as much "out there," independent of his little mind and his culture, as an astronaut exploring the surface of Mars.

In the light of today's physics the entire universe has dissolved into pure mathematics. The cosmos is made of molecules, in turn made of atoms, in turn made of particles which in turn may be made of superstrings. On the pre-atomic level the basic particles and fields are not made of anything. They can be described only as pure mathematical structures. If a photon or quark or superstring isn't made of mathematics, pray tell me what it *is* made of?

To imagine that these awesomely complicated and beautiful patterns are not "out there," independent of you and me, but somehow cobbled by our minds in the way we write poetry and compose music, is surely the ultimate in hubris. "Glory to Man in the highest," sang Swinburne, "for Man is the master of things."

9

Ian Stewart's *Flatterland*

This review first appeared in the *Washington Post*'s *Book World* (June 24, 2001).

Ian Stewart, a mathematician at the University of Warwick in the United Kingdom, is best known here for his column on recreational mathematics in *Scientific American*. But Stewart, who combines a deep understanding of math with an engaging literary style, has written more than sixty books as well. Some of their titles are: *The Magical Maze, Another Fine Math You've Gotten Me Into, Does God Play Dice*?, and *Fearful Symmetry*. *Flatterland* (Perseus, 2001), his latest book, is essentially a work about mathematics cleverly disguised as a fantasy sequel to Edwin Abbott's classic *Flatland*.

Flatland was published in England in 1884 under the pseudonym of A. Square. It describes a two-dimensional world inhabited by male polygons (triangles, squares, hexagons, etc.) and females who are line segments with an optical "eye" at one end like a needle. They slide about over a frictionless surface with no comprehension of a three-dimensional world. A visitor from 3-space, called the Sphere, intersects their plane in the form of a circle. He does his best to convince Flatlanders that there is a larger world of solid shapes, and by implication a still higher reality of four dimensions.

The protagonist of *Flatterland* is Victoria Line, a distant descendant of Albert Square. In the attic of her home she discovers Square's original handwritten manuscript. The book had been censored by Flatland authorities because of its pernicious notion of three dimensions. Victoria is fascinated by the manuscript. At its end, Square added a paragraph in code that Victoria manages to decipher. It tells how to summon the Sphere to Flatland.

After Victoria follows instructions, the Sphere materializes in her bedroom. He turns out to be a Space Hopper—a person with the amazing ability to hop from one space to another. He and Victoria vanish from her bedroom as the Space Hopper takes her on a wild tour of different kinds of spaces. Not only is there an infinity of higher dimensions; there are even fractal spaces with dimensions that are fractions between the integers.

Victoria is a quick study. A diary that she keeps allows Stewart to introduce an awesome variety of dazzling topics that include not only pure mathematics but also the latest speculations of physicists and cosmologists.

The whirlwind tour begins with a visit to what is called Planeturth. (The novel swarms with amusing made-up words taken from *Flatland*'s lexicon and from the Space Hopper's jargon.) There she learns some basic facts about *n*-dimensional geometry, and such curiosities as how a cube can go through a hole in a slightly smaller cube, and Kepler's problem of finding the densest way to pack identical spheres, a problem not solved until 1999.

Victoria's next visit is to the world of fractals and the world of topology, the study of properties unaltered by deformations such as stretching and twisting. Victoria learns about the Moebius strip from a one-sided cow called Moobius; she also meets Alexander's horned sphere, and other topological monstrosities.

Later visits introduce Victoria to projective geometry, graph theory, groups, error correction codes and the hyperbolic plane. The book's last half plunges her into such hairy technical topics as quantum mechanics, where she converses with Schrödinger's cat and learns about relativity, black holes, worm holes, and the Big Bang. Finally, Victoria encounters the most recent candidate for a Theory of Everything: superstrings. These are inconceivably tiny loops that live in a world of ten dimensions. Their different vibrations generate all the basic particles.

Victoria returns home fired with enthusiasm for awakening Flatlanders to the reality of a vast multiverse—a universe of universes—of which they are but an infinitesimal part.

10

Kurt Gödel's Amazing Discovery

This review first appeared in *The New Criterion* (December 2000).

David Hilbert, the great German mathematician (he died in 1943), had a stupendous, dazzling vision. He hoped and believed that some day mathematicians would construct one vast formal deductive system with axioms so powerful that every possible theorem in all of mathematics could be proved true or false. Such a system would have to be both consistent and complete. Consistent means it is impossible to prove both a statement and its negation. Complete means that every statement in the system can be proved true or false.

In 1931, to the astonishment of mathematicians, a shy, reclusive Austrian, Kurt Gödel, age twenty-five, shattered Hilbert's magnificent dream. Gödel showed that any formal system rich enough to include arithmetic and elementary logic could not be both consistent and complete. If consistent it would contain an infinity of true statements that could not be proved by the system's axioms. What is worse, even the consistency of such a system cannot be established by reasoning within the system. "God exists," a mathematician remarked, "because mathematics is consistent, and the devil exists because we can never prove it."

I recall a cartoon by Robert Monkoff which shows a man in a restaurant examining his bill. He is saying to the puzzled waiter: "The arithmetic seems correct, yet I find myself haunted by the idea that the basic axioms on which arithmetic is based might give rise to contradictions that would then invalidate these computations."

Fortunately arithmetic *can* be shown consistent, but only by going outside arithmetic to a larger system. Alas, the larger system can't be proved consistent without going to a still larger system. Many formal systems less complex than arithmetic, such as simple logics and even arithmetic without multiplication and division, can be proved consistent and complete without going beyond the system. But on levels that include all of arithmetic, the need for meta-systems to prove completeness and consistency never ends. There *is* no final formal system, such as Hilbert longed for, that captures all of mathematics. "Truth," as the authors of this new book capsule it, "is larger than proof."

Statements that are true but unprovable inside a formal system are called "Gödel undecidable." Like consistency, they can be shown true by a meta-system, but the larger system is sure to contain its own unprovable statements. There is no escape from the endless regress of meta-systems, each with undecidable theorems.

Many books and thousands of technical papers have dealt with Gödel's epochal bombshell and its implications. Several excellent biographies of Gödel have been published. *Gödel: A Life of Logic*, by American science writer John Casti and mathematician Werner DePauli of the University of Vienna, is a splendid nontechnical account of the Gödelian revolution and at the same time a sketch of Gödel's eccentric life and its tragic ending.

The authors begin with Gödel's birth and childhood in Brno, a city now part of Czechoslovakia, where he became fluent in German, French, and English. In 1924 he settled in Vienna. There his philosophical interests put him in contact with the famous Vienna Circle of philosophers whose leaders were Moritz Schlick and Rudolf Carnap.

Gödel's ingenious method of proving undecidability is explained by John Casti and Werner DePauli, in *Gödel: A Life of Logic* (Perseus, 2000), as clearly as possible in a book for general readers. The argument requires attaching a prime number (a number with no divisors except itself and 1) to each symbol in a sentence expressing a theorem. Multiplying its prime numbers gives a product which, because it has a unique set of prime factors, is a unique coding of the sentence. By applying a "diagonal proof" which Georg Cantor had used to show that the infinity of real numbers exceeds the infinity of whole numbers, Gödel was able to construct a number that codes a true sentence which states its own unprovability. In later decades a raft of simpler ways to obtain the same result have been devised, many by the American logician Raymond Smullyan.

In 1940, now world famous, Gödel joined the staff of eminent thinkers at the Institute for Advanced Study, in Princeton, where he and Einstein became good friends. "I used to see them walk to work together," Murray Gell-Mann recalls in his autobiography. "Did they discuss deep mathematical or physical questions? . . . Or was their conversation mainly about the weather and their health problems?"

Although Gödel continued to make significant contributions to mathematics, as he grew older his speculations became more and more bizarre. In a contribution to *Albert Einstein: Philosopher Scientist*, an anthology edited by Paul Schilpp, Gödel proposed a model of the universe difficult to take seriously. He conjectured that the universe is rotating in such a peculiar way that it generates closed timelike loops along which a spaceship could travel into the past and future. Following Kant, Gödel believed time to be a subjective illusion. The universe itself is timeless; what William James called a "block universe" in which past and future exist in an eternal now. In such a cosmos free will obviously is also an illusion, and it is impossible for anyone to alter either the future or the past.

How, one at once wonders, did Gödel avoid the paradoxes of time travel into the past, such as killing yourself, or performing other acts involving your duplicate—acts not in your memory? Gödel's way out was surprisingly feeble. He argued that travel to the past required such high speeds and long distances that not enough fuel would be available!

Strongly influenced by Leibniz, and by the German philosopher Edmund Husserl, Gödel outlined his own curious version of the ontological argument. This old argument claims to prove that the nonexistence of God, the most perfect of all beings, implies a logical contradiction. Gödel was a philosophical theist who not only believed in a personal God, but also in an afterlife. Because humans have an infinite potential for development, he insisted, it would be absurd for God to create us without allowing for such progress.

As he aged, Gödel became increasingly preoccupied with occultism, reincarnation, and parapsychology. He believed in the reality of ESP and ghosts. He was convinced that his wife, Adele, had strong powers of precognition.[1] He believed that our "self" is more than just the physical brain. "The notion that our ego consists of protein molecules," he said in a letter, "seems to me one of the most ridiculous ever made."

Gödel's philosophy of mathematics, Casti and DePauli make clear, was one of extreme Platonic realism. He thought mathematical objects such as numbers, triangles, and even Cantor's transfinite sets, are as real and independent of human minds, though in a different way, as pebbles and planets.

Bertrand Russell, in the second volume of his autobiography, speaks of discussions he had at Einstein's house with Gödel, whom he mistakenly calls Jewish. "Gödel turned out," Russell writes, "to be an unadulterated Platonist, and apparently believed an eternal 'not' was laid up in heaven, where virtuous logicians hope to meet it hereafter."

Here is how Gödel replied in an unsent letter:

> As far as the passage about me [in Russell's autobiography] is concerned, I have to say *first* (for the sake of truth) that I am not a Jew (even though I don't think this question is of any importance), 2.) that the passage gives the wrong impression that I had many discussions with Russell, which was by no means the case (I remember only one). 3.) Concerning my "unadulterated" Platonism, it is no more "unadulterated" than Russell's own in 1921 when in the *Introduction [to Mathematical Philosophy*] he said "[Logic is concerned with the real world just as truly as zoology, though with its

[1] Adele was an attractive, contentious, antisocial, uneducated woman with atrocious taste. She once planted a pink flamingo outside her Princeton home. Gödel's taste, his biographer John Dawson comments, was no better. He thought the flamingo was "charming."]

more abstract and general features". At that time evidently Russell had met the "not" even in this world, but later on under the influence of Wittgenstein he chose to overlook it.

As Gödel neared death he became increasingly depressed and paranoid. He began refusing food because he thought he was being poisoned. Somehow Adele managed to keep him alive until his weight dropped to sixty pounds and he stopped eating altogether. His death was a suicide brought on by self-inflicted starvation. He left a vast quantity of unpublished papers, and notes kept in a long-forgotten German shorthand called the Gabelsberger method.

Casti and DePauli devote the last half of their book to ramifications of Gödelian undecidability, including the amazing discoveries of IBM's Gregory Chaitin about random numbers, too technical to discuss here. In England, Alan Turing translated Gödel's results into the operation of computers. He showed, using arguments similar to Gödel's, that it is impossible to build a computer that can decide the truth or falsity of every mathematical question. Given a conjecture and a program for testing it, a computer will stop after a finite number of steps if it finds the theorem true or false. But there are undecidable questions it will keep testing forever. The authors are skillful in explaining the "halting problem" as it applies to an idealized computer known as a Turing machine.

Not all undecidable theorems must be true. They can be true or false, and in some cases can be assumed either true or false. For example, consider the notorious parallel postulate given as an axiom by Euclid. It states that through a point outside a straight line one and only one parallel to the line can be drawn. The postulate is undecidable in a system based on Euclid's other axioms. If assumed true, it leads to Euclidean geometry. If assumed false, it leads to systems of non-Euclidean geometry.

Because laws of physics are expressed mathematically, the possibility arises that the universe may also have its undecidable laws. "Some of our colleagues in particle physics," writes physicist Freeman Dyson in *Infinite in All Directions*, "think that they are coming to a complete understanding of the basic laws of nature. . . . But I hope [this] . . . will prove as illusory as the notion of a formal decision process for all of mathematics. If it should turn out that the whole of physical reality can be described by a finite set of equations, I would be disappointed. I would feel that the Creator had been uncharacteristically lacking in imagination."

For decades Fermat's last theorem—the conjecture that $a^n + b^n = c^n$ has no solution if n is an integer greater than 2—was thought by some mathematicians to be undecidable until a few years ago when it was proved true. Today the outstanding unsolved problem in number theory is Goldbach's conjecture. It states that every even number greater than 2 is the sum of two primes in at least one way. All number theorists believe it, partly because it has been tested to monstrously large even numbers, and partly because the larger the number the more ways it can be a sum of two primes. Is it possible that Goldbach's conjecture is undecidable?

Two publishers, England's Faber and Faber and New York's Bloomsbury Books, have offered a million dollars for a proof of Goldbach's conjecture before March 15, 2002. The offer is promotion for a Greek novel by Apostolos Doxiadis with the English title *Uncle Petros and Goldbach's Conjecture.* It tells how Petros Papachristos was driven mad by his vain efforts to prove Goldbach's conjecture before he decided it was undecidable. No proof was found, so the prize was not given.

It is easy to show that if Goldbach's theorem is undecidable it must be true. Assume it undecidable and false. If so, there will be a counterexample, an even number not the sum of two primes, which a computer could find in a finite number of steps. This makes the theorem decidable, thus contradicting the assumption that the conjecture is undecidable and false. Therefore it must be undecidable and true.

Casti and DePauli take up a question which is currently controversial. Does the fact that computers are unable to decide an infinity of conjectures show that our minds are superior to computers because we can often construct meta-systems, with intuitively reasonable axioms, that will decide such questions? This thesis was first strongly defended by Oxford philosopher J. Anthony Lucas, and more recently by Oxford's mathematical physicist Roger Penrose and others. Almost all artificial-intelligence experts disagree. They are convinced that computers of the kind we know how to build—that is, computers made of wires and switches—are capable in principle of doing as well as humans in finding meta-systems. Indeed, some experts believe that computers will soon cross a threshold of complexity that will enable them to do everything human minds can do.

Physicist Jeremy Bernstein, in one of his books, recalls that as a student he once asked the great Hungarian mathematician John von Neumann, a friend and admirer of Gödel, if computers would ever replace mathematicians.

"Sonny," von Neumann replied, "don't worry about it."

11

New Results on Magic Hexagrams

This article first appeared in *The College Mathematics Journal* (September 2000).

Combinatorial problems involving magic squares, stars, and other geometrical structures often can be solved by brute-force computer programs that simply explore all possible permutations of numbers. When the number of permutations is too large for a feasible running time, an algorithm can frequently be reduced to manageable time by finding ingenious shortcuts. Such planning makes computer solving less trivial and much more interesting

A superb example of such planning was described in a little-known short article in the *Mathematical Gazette* (Vol. 75, June 1991, pp. 140–42). The authors, Brian Bolt, Roger Eggleton, and Joe Gilks, posed for the first time a problem based on the pattern shown in Figure 1. It is the traditional hexagram or Star of David with its inner hexagon divided into six equilateral triangles. Can numbers 1 through 12 be placed inside the twelve triangles so that each of the six rows, indicated by the arrows, has the same sum?

The authors point out that there are 12! ways of arranging the numbers. When the pattern's six rotations, and six reflections, are excluded,

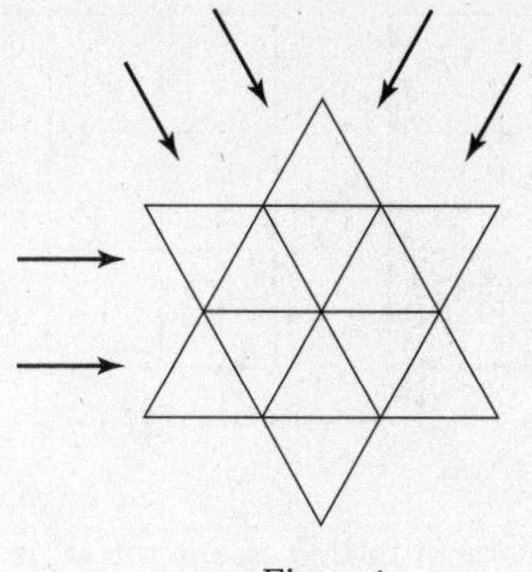

Figure 1

the number reduces to 11! or 39,916,800. Can this number be further reduced?

It can. Each of the six rows must have the same sum, and this "magic constant" must be either odd or even. It is not hard to discover that, by neglecting rotations, reflections, and complements, there are just three ways that odd and even numbers from 1 to 12 can be distributed on the hexagram (see Figure 2).

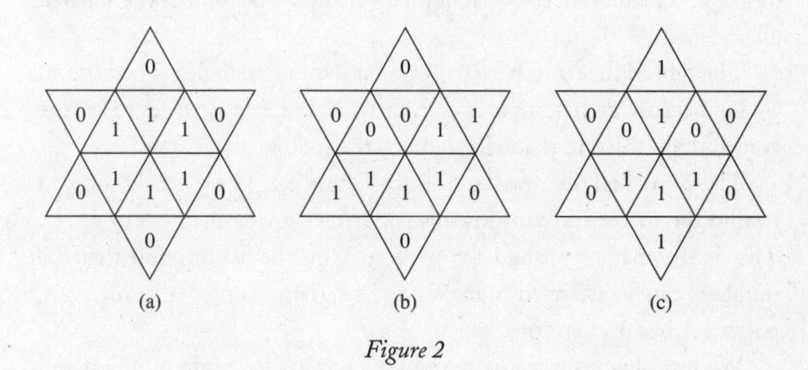

Figure 2

A complement, I should explain, is obtained by subtracting each number of a magic square or star from the pattern's largest number plus 1. For example, consider the Chinese *lo shun* or 3 × 3 magic square in Figure 3. If each number is replaced by the remainder when it is taken from 10, we obtain the complement (Figure 4).

8	1	6
3	5	7
4	9	2

Figure 3

2	9	4
7	5	3
6	1	8

Figure 4

Note that the complement in this case is the same as before except for a rotation and reflection. In the case of magic hexagrams, complements are always different, but are considered trivial variations. The complements of the three patterns shown in Figure 2 are obtained by replacing each 1 with 0 and each 0 with 1. Each 1 stands for an odd number, each 0 for an even number. By neglecting complements, as well as rotations and reflections, the three odd-even patterns provide a way of establishing upper and lower bounds for the star's magic constant.

In pattern A all the even numbers go in outside triangles. They add to 42, but because each number appears in two rows, we double 42 to get 84 as the contribution the even numbers make to the total of the sums of all six rows.

The odd numbers in pattern A go on interior triangles. They add to 36, but because each number is in three rows we triple 36 to get 108 as the contribution the odd numbers make to the total of all six rows.

The sum of all the rows' sums is 84 + 108 = 192. There are six rows, so to find the magic constant for this pattern we divide 192 by 6 to get 32. This is the magic constant for pattern A on the assumption that the numbers can be arranged to make the hexagram magic. As it turns out, pattern A has no solution.

We turn now to patterns B and C. Each has four odd numbers and two even numbers on the interior triangles. To make the magic constant as low as possible, we place on these inside triangles, where each number appears in three rows, the four lowest odd numbers (1, 3, 5, 7) and the two lowest even numbers (2, 4). The sum of the inside numbers is 22. Tripling it gives 66.

On the outside triangles go odd numbers 9, 11 and even numbers 6, 8,

10, 12. They add, to 56. Each number appears in two rows, so we double 56 to get 112. Adding 66 to 112 gives 178.

We know that this total must be a multiple of 6 because six rows must have the same sum. So we raise 178 by the smallest amount to arrive at a multiple of 6, namely 180. Dividing 180 by 6 yields 30. This is the lower bound for the magic constants of patterns B and C.

To obtain the upper bound we put the largest two even numbers (10, 12) and the four largest odd numbers (5, 7, 9, 11) on the interior triangles. They add to 54 and three times 54 is 162. The outside numbers (1, 2, 3, 4, 6, 8) add to 24, and twice 24 is 48. The sum of 162 and 48 is 210. Dividing 210 by 6 gives 35 as the upper bound for the magic constants of patterns B and C.

We now have shown that the magic constant, *if* such a magic hexagram does indeed exist, must be 30, 31, 32, 33, 34, or 35. Knowing this, and knowing how odd and even numbers must be distributed to make the hexagrams magic, allowed the three authors to reduce their computer program to a manageable length.

To the authors' vast surprise, their computer search produced just one solution! It is shown on the left of Figure 5. Its complement (each number taken from 13) is on the right. "When we discovered the beautiful solution," the authors write, "we rushed from our baths into the street shouting 'Eureka!' " The two patterns make excellent puzzles. Place numbers 1 through 12 on the hexagram so each row adds up to 33, or so each row adds up to 32. In both cases the solution is unique. Note that in the pattern on the left each pair of opposite corners adds up to 12, and on the right, they total 14.

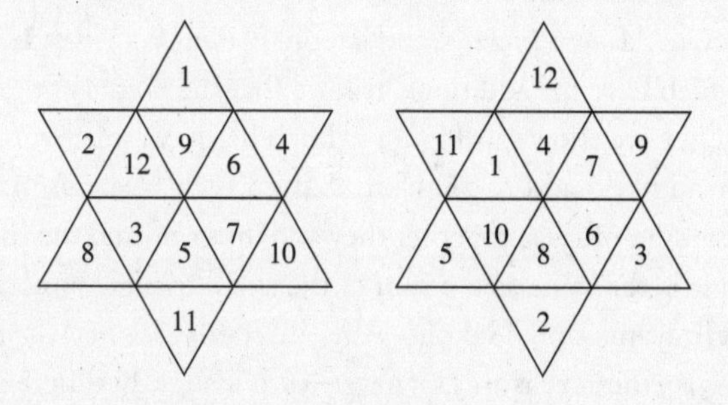

Figure 5

A much older magic hexagram problem, now thoroughly explored, consists of placing numbers 1 through 12 on the vertices of the traditional Star of David shown in Figure 6. Because each number appears in two rows, the sum of all the row sums is twice the sum of numbers 1 through 12, or 78 × 2 = 156. There are six rows, so 156/6 = 26, the hexagram's magic constant.

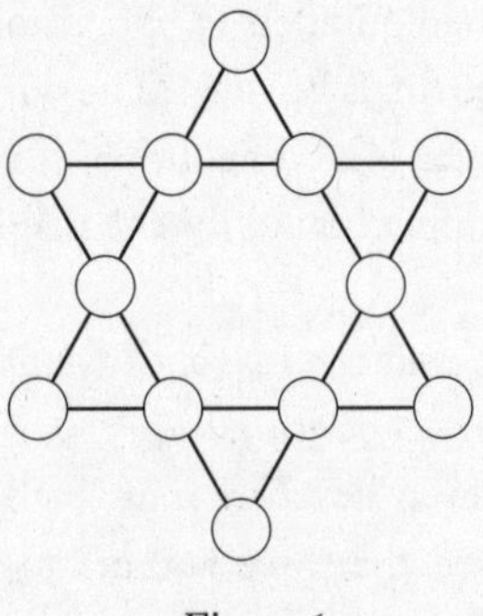

Figure 6

The British puzzlist Henry Ernest Dudeney, in *Modern Puzzles* (1926) claimed there are 37 fundamental solutions, or 74 if complements are included. This was one of Dudeney's rare mistakes. He missed three basic patterns. Von J. Christian Thiel, in the German periodical *Archimedes* (vol. 5, September 1963, pp. 65–72) displays fifteen basic solutions. By applying well-known transformations in which numbers are interchanged a certain way, Thiel raises the total of patterns, not counting rotations and reflections, to forty. These, together with their complements, make eighty solutions.

Dudeney, in *Modern Puzzles*, and later in *A Puzzle-Mine*, identifies six solutions which have the additional feature that the points of the star also add to 26. They are shown in Figure 7. Each has a complement in which the interior numbers add to 26. Note that on each hexagram each large triangle has a sum of 13, so together they add up to 26, and the sum of any small triangle is the same as the sum of the small triangle opposite.

Akira Hirayama and Gakuho Abe, in *Researches in Magic Squares* (Osaka, 1983)—the text is in Japanese—give a slightly different way of arriving at eighty fundamental solutions. The book also includes Abe's

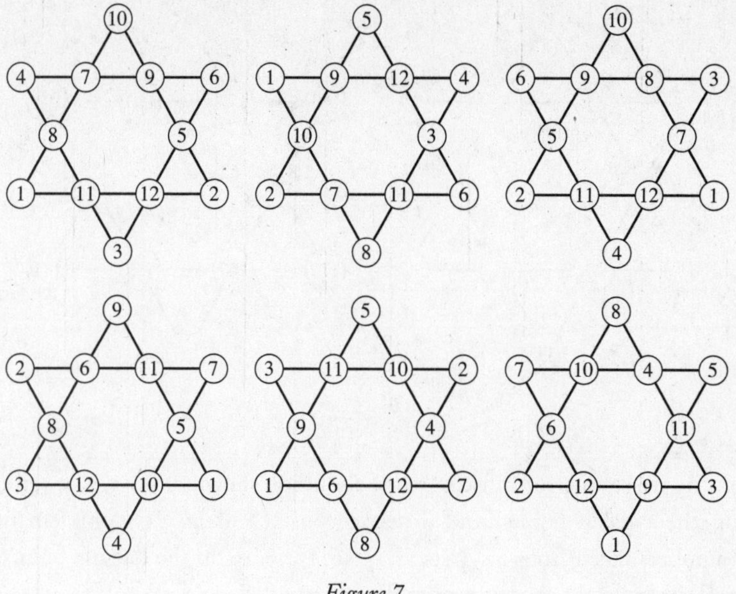

Figure 7

amazing constructions of two magic hexagons made with consecutive primes! They are shown in Figure 8.

There is now a sizeable literature on stars with more than six points. The five-pointed star, or pentagram, is easily shown to have no solution for numbers 1 through 10. You'll find one such proof in the chapter on magic stars in my *Mathematical Carnival.* In the same chapter I explain how the numbers in a solution for the traditional hexagram can be transferred to the twelve edges of the cube so that four numbers around each face total 26. Because the cube's "dual" is the octahedron (faces and vertices exchanged), the same numbers can be placed on the octahedron's edges so that the four edges around each vertex add to 26. Figure 9 shows how a hexagram solution is transferred to a magic cube and to a magic octahedron.

A third type of hexagram is formed by adding three diagonals to the traditional structure as shown in Figure 10. The question arises: Can numbers 1 through 19 be placed on the vertices of this pattern so every line of five has the same sum? Harold Reiter, at the University of North Carolina, Charlotte, posed this as an unsolved problem in his article "A Magic Pentagram," in *Mathematics Teacher* (March 1983, pp. 174–77).

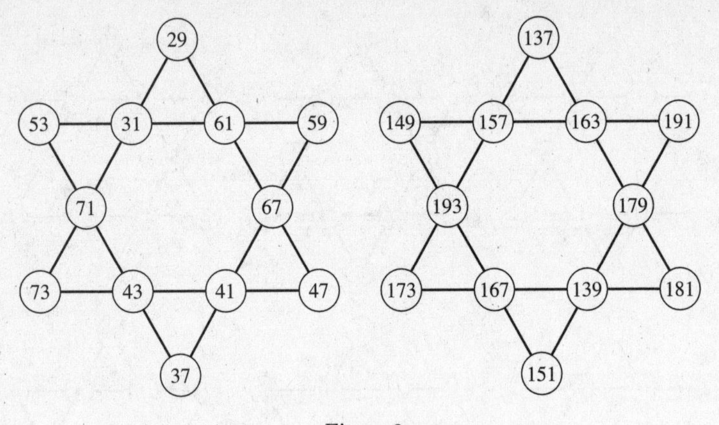

Figure 8

By trial and error I discovered the solution shown; it has 1, 2, 3, 4, 5, 6 on the outside points, and a magic constant of 46. Its complement (numbers taken from 20) puts 14, 15, 16, 17, 18, 19 on the outside points, and raises the magic constant to 54.

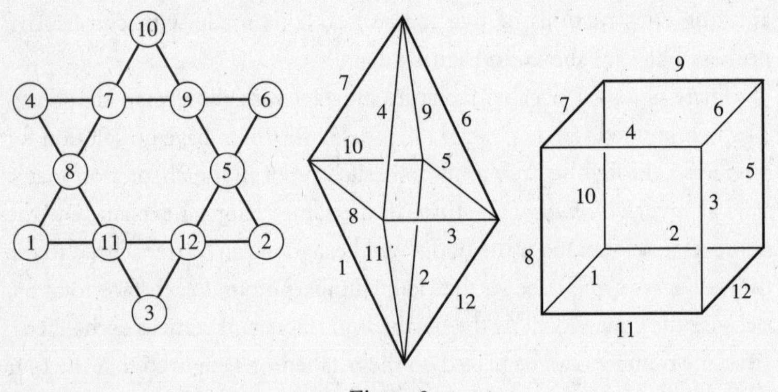

Figure 9

Finding all solutions to this problem by a brute-force computer algorithm is out of the question because it would require examining 19!/12 possible permutations. Reiter and his associate David Ritchie, in "A Complete Solution to the Magic Hexagram Problem," in *The College Mathematics Journal* (September 1989), describe the clever shortcuts they used to reduce their algorithm to an examination of a mere 18,264,704 patterns. In less

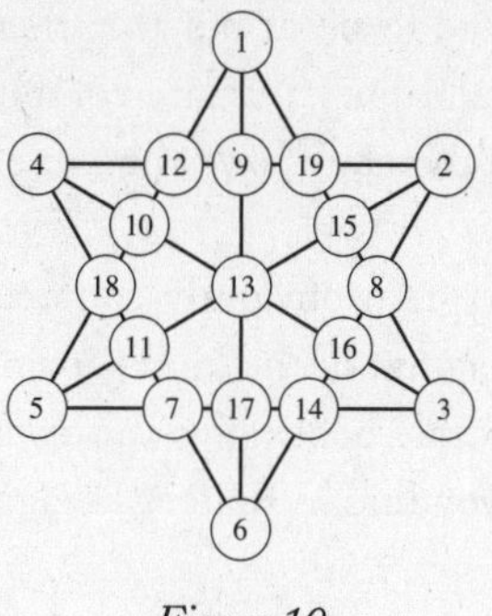

Figure 10

than five minutes their Pascal program found 2,190 basic solutions, not counting rotations, reflections, and complements, or 4,380 solutions if complements are included. The magic constants range from 46 through 54.

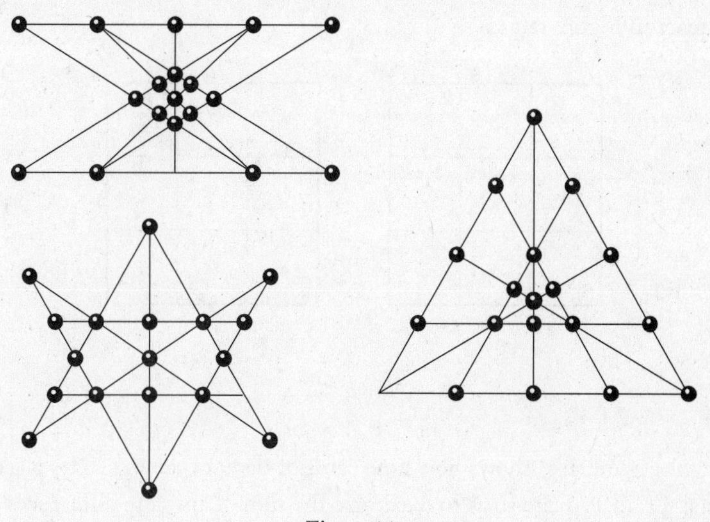

Figure 11

The hexagram pattern in Reiter's problems is an interesting one. It solves an old tree-plant puzzle of placing nineteen trees to form six straight rows with four trees in each row. I used this pattern for a checker-like game I named Solomon. Its board and pieces are available from Kadon Enterprises, 1227 Lorere Drive, Pasadena, Maryland 21122.

Frank Bernhart wrote to point out that there are many symmetric patterns with vertices and lines that are combinatorially equivalent to Reiter's hexagram, and therefore have the same set of solutions. Three are shown in Figure 11.

The traditional hexagram obviously contains eight different triangles. The modified hexagram of the kind shown in Figure 1 has twenty triangles. (Many old puzzle books give this as a problem.) How many different triangles can you find in Reiter's hexagram? Counting them all is not so easy.

Answers:

1. Move the bottom two of three cards to the top, or move the leftmost column to the right. Either change produces the desired magic square in Figure 12(a).

2	3	A
A	2	3
3	A	2

(a)

A	K	Z
Q	Y	2
X	3	J

(b)

Figure 12

2. Figure 12(b) shows how nine cards of distinct values can be placed in a toroidal square so as to maximize the sum of absolute differences of adjacent values in rows and columns. Any value 4 through 10 can go in cells, X, Y, and Z to make a total of 120. This was proved maximal by Brian Maxwell, of Middlesex, England, in the *Journal of Recreational Mathematics*, Vol. 18 (1985–86), p. 300.

PART III

Religion

12

The Strange Case of Garry Wills

A much truncated version of this review appeared
in the *Los Angeles Times* (August 6, 2000).

Garry Wills is a Roman Catholic scholar of awesome erudition whose more than twenty books on politics, religion, and other topics are models of brilliant rhetoric and beautiful writing. He seems to have read everything even remotely relevant to any topic. *Papal Sin* (Doubleday, 2000), his latest work, will surely be the most controversial. It will generate loud cheers from Protestants, Jews, Muslims, philosophical theists, and even atheists. The philosopher Richard Rorty, a secular humanist, had high praise for the book in the *New York Times Book Review* of June 11, 2000.

Reactions by Catholics will of course be mixed. My guess is that those as ultra-orthodox as Bill Buckley and Pat Buchanan will be appalled. Catholics as liberal as Father Greeley will praise *Papal Sin* as a courageous effort to reform the faith which they each love and know so well. At the end of this chapter I will explain why, though I agree with all of Wills's passionate attacks on certain popes, I find something mysterious and strange about him.

Born in 1934 in Atlanta, with a doctorate from Yale (his thesis was on

Aeschylus), Wills is currently an adjunct professor of history at Northwestern University. His writing career began at age twenty-three, when after obtaining a master's degree in 1958 at Xavier, a Jesuit university. He was hired to write book reviews and drama criticism for Buckley's *National Review*. In *Confessions of a Conservative* (1979) Wills recalls being asked by Buckley, during their first meeting, if he had left the Church. Buckley was relieved when Wills said no. "Being Catholic," Wills adds, "always mattered more to him than being conservative."

At that time Wills called himself a distributist, G. K. Chesterton's term for a movement in England which favored the redistribution of wealth from rich to poor. Since he left the *National Review*, Wills has distanced himself from Buckley by steadily moving leftward both politically and religiously.

Wills's lack of respect for recent popes surfaced early in his career, notably in *Bare Ruined Choirs* (1972). In this vigorous attack on the Church's heirarchy, Wills likens the Vatican to the "bare ruin'd choirs where late the sweet birds sang" in Shakespeare's Sonnet Seventy-three. Rome is faulted for its stubborn opposition to contraception, ordination of women, and the marriage of priests. Adjectives which Wills hurls at the Church include cracking, weakening, disintegrating, and crumbling. As a loyal practicing Catholic, he urges his Church to cast off its shackles and welcome history's irreversible changes.

Papal Sin continues the themes of *Bare Ruined Choirs* with greater fury and more sordid details. In some ways Wills resembles Martin Luther nailing his charges on a Church door. In other ways his book follows the long tradition of thousands of books, mostly by Protestants and ex-Catholics of past centuries, that trash the lives and beliefs of earlier popes.[1]

The first four chapters of *Papal Sin* are scathing attacks on the Church, not just for its past anti-Semitism and its horrendous pogroms, but for its continued efforts to cover up this awful history. Not until 1985 did Rome officially repudiate the claim that Jews are under a special

[1] Recent examples of Pope bashing include E. R. Chamberlin, *The Bad Popes*; Peter De Rosa, *Vicars of Christ*; Frederick Ide, *Unzipped: The Popes Bare All*; and Simon Duffy, *Saints and Sinners: A History of the Popes.*

curse from God for having executed and repudiated their own Messiah. "Seminaries taught it," Wills writes, "and Biblical commentaries explained it, and persecutions were based on it."

Throughout the Second World War and after, Pope Pius XII privately helped many Jews escape from Germany, but not once did he speak out against Hitler or the holocaust. He may have imagined he had good reasons for this silence, but this is no excuse, Wills argues, for the Vatican to pretend now that the Church did not swarm with European bishops and priests who supported the Nazis. Not until 1988 did the Vatican issue *We Remember*, a document which Wills sees as a dishonest effort to minimize with fake history the feebleness of Rome's response to Hitler's madness.

Chapter 3 tells the tragic story of Edith Stein, a German philosopher who converted to Catholicism, became a nun, and died at Auschwitz. The Vatican maintains, against all evidence, that Stein was arrested because she was a Catholic, not because she was a Jew. John Paul II made her a saint and martyr in 1998. Only one miracle is required for martyrdom. In Stein's case it was the recovery of a two-year-old girl from an overdose of Tylenol. Because almost all Tylenol overdosers recover, why did John Paul deem this a miracle? Because the girl's parents had prayed to Edith Stein!

Wills scoffs at this. He sees Stein's canonization as little more than an attempt to spread the false notion that many Catholics were also Hitler's victims. There was a second similar effort. A Polish priest, well known for his anti-Semitism, also died in a death camp. He, too, was canonized by John Paul.

Chapters 5 and 6 excoriate the Church for its perpetual ban on all forms of birth control except the undependable and degrading rhythm method. Wills reminds us of the dreadful attitude toward sex that prevailed throughout the Church's early history. The Fathers and St. Augustine taught that sex was always sinful unless the act was strictly for the purpose of producing a child. Augustine would have considered the rhythm technique a stark violation of this rule. St. Thomas Aquinas thought sex immoral unless the man was on top, and that using a contraceptive was worse than incest!

Paul VI's encyclical letter *Humanae Vitae* (1965) prohibited all forms of contraception, including the pill. Wills calls it the "most disastrous papal document of the century. . . . the most crippling, puzzling blow to organized Catholicism." The prohibition was instantly rejected by the overwhelming majority of Catholics throughout the world. "Sex is for procreation, yes—," Wills writes, "but all the time, at each and every act? Eating is for subsistence. But any food or drink beyond that necessary for sheer subsistence is not considered mortally sinful."

In his next two chapters Wills turns his anger toward the Church for refusing to ordain women. Grounds for the ban are in Wills's eyes preposterous: The apostles were men, women don't look like Jesus, Paul's letters tell women not to speak in church, and so on. It's hard to believe that the following passage, which Wills quotes from Aquinas, was ever written:

> In terms of nature's own operation, a woman is inferior and a mistake. The agent cause that is in the male seed tries to produce something complete in itself, a male in gender. But when a female is produced, this is because the agent cause is thwarted, either because of the unsuitability of the receiving matter [of the mother] itself or because of some deforming interference, as from south winds, that are too wet, as we read in [Aristotle's] *Animal Conception* (ST 1 q 92, 1 ad 1).

Such fantastic superstitions can be excused, Wills says, on grounds of ignorance, but today, when no one doubts the equality of men and women, the refusal to ordain women by invoking "weird and Scripturally simpleminded arguments" is a "modern sin, and it is a papal sin."

In the next two chapters Wills aims his guns on the Vatican's refusal to let priests marry. The result has been disastrous. Hundreds of priests have secret mistresses. An enormous number have fled the Church to marry. They have been replaced by an increasing number of gays—20 percent according to recent surveys, half of them sexually active. Over 80 percent of priests admit they regularly masturbate.

Not until the twelfth century, Wills reminds us, was celibacy enforced. Earlier priests and bishops had wives. Even Peter, whom the Church considers its first pope, was married, as were other apostles. The Church insists that a priest's holy duties would be severely distracted by marital obligations. Wills responds: "Do any of us feel we must find an unmarried doctor, since no other will be able to give our health his full attention?"

Chapter 11 hammers the Church for its harsh attitude toward divorce, forcing good Catholics to risk damnation by obtaining a civil divorce and living in mortal sin with a new spouse. In recent years this cruelty has been mitigated by annulments. In the United States alone, Wills reports, six thousand annulments are now granted annually.

The next two chapters survey the Church's baleful record on homosexuality, and its strenuous efforts to conceal the number of priests who were and are compulsive pedophiles. Wills tells several horror tales about such cases. Instead of having the priest arrested, he was merely shifted to another parish and his crimes hushed up. As for gays and lesbians who lead exemplary lives, Wills sees no reason why they should not be priests and Protestant ministers.

Wills moves closer to heresy in Chapter 14—a chapter about Mary. Mariolatry, an excessive devotion to the virgin, he informs us, did not get under way until the sixth century. Augustine preached hundreds of sermons, not one about Mary. Aquinas strongly opposed the doctrine of immaculate conception (that Mary was without original sin), even quoting gospel passages in which Jesus treats his mother with curious indifference. Wills cites the Lord's disrespectful remarks to his mother at the Cana wedding feast. Mariolatry, he writes, reached its peak on the only two occasions when a pope spoke infallibly: to proclaim true the doctrines of the immaculate conception (1854) and the assumption of Mary into heaven (1950).[2]

On pages 174 and 219 of Wills's book, he makes a startling suggestion. He believes that many of the functions now assigned to Mary

[2] Not until the fifth century did the belief become widespread that three days after Mary died her body was taken to heaven to reunite with her soul. The legend varied, but most versions tell that when a Jew tried to upset the bier an angel punished him by cutting off his hands. Peter later restored them and the Jew became a believer. Other Jews who tried to burn the body were stricken with blindness. To convince doubting Thomas of her assumption, Mary dropped her girdle down from the skies. Aquinas accepted the assumption as fact, and so did Cardinal Newman.

should be transferred to the Holy Spirit. He would like to see the Holy Spirit called She, not It. He hopes that prayers now directed to Mary would be directed to Her.

Although Wills thinks that every effort should be made to avoid abortion, he attributes (in chapter 15) its widespread practice to the Church's policy on contraception. Lifting this ban, he contends, would be the most effective antiabortion plan the Church could adopt.

The next three chapters are devoted to three of Wills's heroes: Lord Acton, Cardinal John Newman, and Saint Augustine. When Acton, a devout Catholic historian, made his famous remark about absolute power corrupting absolutely, he was referring, Wills reveals, not to political leaders but to popes. Newman is lauded for his opposition to papal arrogance. Augustine, about whom Wills has written a splendid book, is a saint he admires far more than Aquinas.

Wills's final chapter is surprising and none too clear. He praises René Girard, who had been his colleague at Johns Hopkins University before Wills was fired for starting a column in a Catholic magazine, and for his rejection of the doctrine of the atonement. Jesus did not die as a sacrifice for our sins. The true meaning of his death is its final triumph over Satan. Wills writes:

> Jesus lets the violence of the world system defeat itself on his dying body—instead of this being a sacrifice to a vengeful God, it is a paradoxical defeat of the torturer. The fallen world of satanic resistance to God causes the final violence, not any placatory act demanded by the Father. The only sacrifice by Jesus is his offering of his innocent body to the fury of the sacrificial system that is being canceled. This was exactly the position of Augustine. In an early work, he opposed the ransom theory of Christ's death, the theory that Jesus was a substitute who accepted the suffering that the Father wanted to inflict on others—as if the Father could find satisfaction in causing pain: "The Lord's was obviously not a death of ransom but of restoration (*dignitatis non debiti*)."

We don't know, of course, if these are Wills's sentiments or if he is merely reporting what his friend and Augustine believed about the purpose of the crucifixion. If Wills has indeed abandoned the atonement doctrine, this seems to me another instance of his constant flirtations with heresy.

Wills closes his book by pleading for his Church to listen more carefully to the Holy Spirit, to renounce all falsehoods, to cease regarding the pope as an emperor and Mary as an empress. The real Church, he insists, is not centered on the Vatican but on wherever the Holy Spirit guides those who minister to the sick and poor, who oppose capital punishment and weapons of mass destruction.

> I do not think that my church has a monopoly on the Spirit, which breathes where She will, in every Christian sect and denomination. In fact, She breathes through all religious life, wherever the divine call is heeded, among Jews and Buddhists and Muslims and others. But we Christians believe She has a special role to complete Christ's mission in us. Unworthy as we are, She calls us. She even calls the Vatican. All Christians need to respond to that soliciting. Including Popes.

This tolerance of other faiths is certainly not characteristic of orthodox Catholics, who are sure that their Church is the Creator's only true revelation. Wills's words, though I find them admirable, sound more like the words of a liberal Protestant or a philosophical theist. Note that he writes "we Christians," not "we Catholics." Is he suggesting, perhaps unconsciously, that he is a Catholic only in some vague nominal sense?

In reviewing ten books (one of them Buckley's recent confessional, *Nearer My God*), in *The New York Review of Books* (February 19, 1998), Wills reveals that he prays "to the Blessed Virgin every day, often using the rosary." Here is how he ends the review: "Let me repeat that I write this as a Marian devotee, though not as a Mariolater. I also write as a Papist. A Papist is not necessarily a papalist. Even a pope need not be a papalist. Just look at John XXIII."

Aside from Protestant fundamentalists, who are never ashamed of telling us exactly what they believe about anything, there is a curious reluctance these days for Christians, both Protestant and Catholic, to let others in on their fundamental convictions. In his introduction to *Heretics*, Chesterton says that the most important thing to know about any person is that person's deepest beliefs. With many eminent Christians of recent times these are not easy to find out.

Did Reinhold Niebuhr believe in life after death? Nobody knows. Did Norman Vincent Peale believe in the Virgin Birth? Exactly what doctrines did Mortimer Adler accept when he became a Catholic convert in 1999, a year before his death. In his prior Episcopalian phase he wrote an entire book about angels with nary a hint anywhere in the book whether he did or did not believe angels actually exist!

Father Frederick Copleston wrote a marvelous multivolumed history of philosophy. I have no inkling of what he believed about any Catholic doctrine. Does Father Greeley accept the immaculate conception and assumption of Mary? He has written movingly about Mary, but without telling us.

I could go on for pages about the doctrinal fuzziness of leading Christian theologians and writers of recent decades. We know that Wills prays to Mary and regularly attends mass. We know how much he loves his Church and wants to revitalize it. Friends constantly urge him to leave Rome, but there is no indication he ever will. "I'm still a loyal Catholic," he once declared. "As Phil Berrigan says, 'The Church is a whore, but she's our mother.' "

Father Hans Küng, the famous German theologian, has never hesitated to make clear that he disbelieves every doctrine unique to Catholicism, yet he has opted to remain a Catholic priest. Had he lived two centuries ago he would surely have been excommunicated. Unlike Küng, Wills is coy about core beliefs. It seems obvious he doubts such doctrines as papal infallibility, the immaculate conception, the assumption of Mary, and the mystical presence of Christ's body and blood in the bread and wine of the Eucharist. Yet he chooses to remain silent on a raft of other doctrines even more essential to historic Christianity. One longs to know how he would answer the following questions:

Do you believe that:

1. Jesus rose bodily from the dead?
2. The Lord walked on water, multiplied bread and fishes, turned water to wine, cast devils into pigs, and brought back to life the corpse of Lazarus and others?

3. Jesus was born of a virgin and had no human father?
4. Satan is a real person?
5. Angels and demons exist?
6. Some sinners will suffer forever in hell?

Wills and I share an admiration for Chesterton. Wills's first book was a biography of G.K., and I recently annotated Chesterton's masterpiece, *The Man Who Was Thursday.* If G.K. had been handed my list of questions I think he would have readily answered. Augustine and Aquinas answered all of them affirmatively in their writings. If Wills, like Küng, no longer believes any of the doctrines that are distinctively Catholic, and especially if he would answer no to my six questions, one is entitled to wonder why he does not walk out of the Church and declare himself a liberal Protestant or a philosophical theist. This is the mystery and strangeness that hovers like a gray fog over everything Wills has written about his faith.

ADDENDUM

As I expected, Catholic periodicals reacted harshly to Wills's book. Robert Royal, for example, reviewing the book for Buckley's *National Review,* considered it a monstrous product of Wills's hubris.

Apparently I, a philosophical theist unattached to any organized religion, take the Catholic faith more seriously than Wills. It is either what it claims to be, God's one true revelation to humanity, or it isn't. I can only conclude that Wills clings to Rome not so much in the dim hope of reforming the Church, but mainly for sentimental reasons. It revives memories of a happy Catholic childhood.

We may soon know his reasons. Wills's next book, scheduled for publication in October 2002, is titled *Why I Am a Catholic.*

I have not yet had the opportunity to read Wills's new book, but I assume he wants to remain in the Church he loves where he can be most effective in altering it in radical ways that he deems desirable.

13

Mad Messiahs

This essay first appeared in *The Skeptical Inquirer* (March/April 2000).

The word *Christ* is Greek for Messiah. When the woman at the well asked Jesus if he was indeed the Messiah, he replied, "I that speak unto thee am he" (John 4:26). Christians have always assumed he was exactly that. Jews of the time obviously could not agree. After all, Jesus failed to restore Jewish power and to bring about a world of peace and justice.

Over the centuries following Christ's death, scores of Jews in Europe either claimed to be the promised Messiah, or were so regarded by their followers. The history of these false claimants is well summarized in Albert Hyamson's seven-page article "Messiahs (Pseudo-)," in James Hastings's *Encyclopaedia of Religion and Ethics*, Volume 8. Here I shall be concerned mainly with Sabbatai Sebi (his name is also spelled Shabbetai Zevi), the most famous of such claimants. He was responsible for the largest messianic movement in Jewish history.

Sabbatai was born in Smyrna, the former name of the Turkish port of Izmir, the son of a Spanish employee of a British mercantile firm. The exact date of his birth is unknown, but was probably 1626. As a youth he was said to be unusually handsome and intelligent, with a fine singing voice.

There was widespread belief among Jews of the time (when they were under harsh persecution) that the Messiah would soon appear and restore them to their homeland. Young Sabbatai immersed himself in the Cabala and its complex numerology. Encouraged by his father, he slowly persuaded himself that he was none other than the anticipated Anointed One. It was not long before his vigorous preaching won him a large following of fervent disciples known as Sabbateans.

In the 1650s the rabbis of Turkey branded Sabbatai a heretic, excommunicated him, and booted him out of Smyrna. In Salonica, where he continued preaching, he met similar resistance and was expelled from that city also. Such expulsions only increased the number and enthusiasm of his followers. Two early marriages ended in divorce before Sarah, a Polish woman of great beauty, announced that God had told her she was to be the Messiah's bride. Sabbatai sent for her, and they were married in Cairo in 1664.

After several years of wandering about the Orient, preaching and gaining new converts, Sabbatai returned to Smyrna in 1665 where he openly proclaimed himself the Messiah and was welcomed with high enthusiasm. About half of all orthodox Jews around the world were now convinced that Sabbatai was indeed the King of the Jews who would lead them back to their Promised Land. One of his top devotees, Nathan Levi of Gaza, assumed the role of Elijah by proclaiming that Sabbatai was Israel's true savior and that 1666 marked the beginning of the Messianic Age.

The tale now takes an astonishing, comic turn. In Constantinople, in 1666, Sabbatai was arrested by the Sultan of Turkey, put in chains, and sentenced to prison. The Sultan gave him a choice of either being executed or converting to Islam. Sabbatai and his wife at once converted!

Sabbatai's tens of thousands of followers were of course crushed. As the *Encyclopaedia Britannica*'s eleventh edition puts it, the Sabbateans reacted with "a sense of shame joined to feelings of despair." The false Messiah's books and letters were burned. His name was so thoroughly erased from Jewish history that it was almost as if he never existed. Scholars today who are acquainted with his strange life believe he was a seriously deranged manic-depressive of great charisma during his manic phases, and given to frequent bizarre behavior.

After Sabbatai's conversion to Islam, he took the name of Aziz Mehmed Effendi, and founded Dönmeh (Turkish for "apostates"), a weird Muslim cult that combined Islamic theology with Jewish beliefs and rituals. After Sarah died in 1674, Sabbatai married a woman named Esther. Eventually he was banished to Dulcigno, now in Yugoslavia, where he died in 1676.

A small number of Sabbateans continued to honor his memory. The last of their leaders was Jankiev Lebowicz, a Polish charlatan who took the name Jacob Frank. Born in 1726, the son of a Polish rabbi, he claimed to be the reincarnation of Sabbatai. Eventually he and his followers, known as Frankists, converted to Russian orthodox catholicism! In 1760 Frank was arrested in Warsaw on a charge of heresy, and spent thirteen years in prison. After his release he called himself the German Baron of Offenbach. According to Hastings's *Encyclopaedia*, "He lived in state until his death in 1791 . . . in various continental capitals, always with an immense retinue and a vast treasure derived from his infatuated adherents." After his death his daughter Eva headed the Frankist cult as its "holy mistress" until she died in 1816. Not until the early 1900s did the cult finally totally expire.

There have been many false messiahs since, but none who gained the widespread devotion of the man later called The Mad Messiah. For details about his life see the article in Hastings's *Encyclopaedia* previously cited, and the thirty pages on him in *The Jewish Encyclopedia*. See also its seventeen pages on Jacob Frank. A German book on Sabbatai, by Josef Kastein, was translated and published in 1931 by Viking with the title *Messiah of Ismir*. Israel Zangwill wrote a brilliant fictionalized account of Sabbatai's life in *Dreamers of the Ghetto* (1948). On August 10, 1997, and in later reruns, the TV series *Mysteries of the Bible* featured Sabbatai's deranged career.

What is the current situation regarding the Second Coming of Jesus and the Jewish messianic hope? Both Judaism and Christianity are now divided into three main groups. In Christendom, fundamentalists continue to expect to be raptured into the skies any day now, before the world as we know it approaches its demise. Conservative Christians, Protestant and Catholic, follow Saint Augustine in moving the time of the Second Coming so far into the future, and so vague in meaning, that they seldom think about it. Liberal Christians long ago took the Second Advent to be no more than a symbol of the world's gradual progress. Julia Ward Howe, a Unitarian minister, opened her great "Battle Hymn of the Republic" with "Mine eyes have seen the glory of the coming of the Lord." She did not mean a literal return of Jesus. For her the Civil War, by abolishing the evil of slavery, was a major step in progress toward a saner world. Onward Christian soldiers!

Religious Jews fall into three similar camps. The orthodox follow Maimonides, who wrote, "I believe in perfect faith in the coming of the Messiah, and though he tarry, I will wait daily for his coming." Conservative Jews have pushed the date far into the future and are as vague about its significance as conservative Christians are about the return of Jesus. Reform Jews regard the Messiah as no more than a symbol of hope for a world free of war and major injustices. Now that Jews have restored their homeland and regained political power, it seems unlikely that demented claimants like Sabbatai will flourish once more to flimflam the faithful.

Nevertheless, in spite of the endless false messiahs in the past, ultra-orthodox Jews around the world periodically declare that the true Messiah is either here or on his way here to work great wonders. They saw the collapse of Soviet Communism as a great sign that the promised redeemer was already on the scene.

The latest instance of such frenzy broke out in Israel in 1992. An ultra-orthodox group called the Habad suddenly decided that the Messiah was about to declare himself. They put up large billboard signs all over Israel saying "Prepare for the Coming of the Messiah." Similar messages appeared on bumper stickers, and on electric signs atop cars. A full-page ad in the *New York Times* was titled "The time for your redemption has arrived."

The Habad is a branch of the Lubavitch movement, named after the Russian village in Smolensk where it began in the eighteenth century. Lubavitchians in turn are part of Hasidism. Most Habad/Lubavitch/Hasidism followers live in Israel and in New York City.

And who was the Messiah that the Habadniks claimed was on the verge of revealing his identity? None other than Menachem Mendel Schneerson (1902–1994), then an eighty-nine-year-old rabbi living in

Brooklyn! Schneerson came from a family that for two centuries had been leaders of the Habad movement. There are some two hundred thousand such believers worldwide, with thirty thousand living in New York City where Schneerson was their world leader.

Rabbi Schneerson (Corbis)

Rabbi Schneerson never claimed to be the Messiah, but his incessant preaching that the Messianic Age was about to begin, together with feeble disclaimers that he wasn't the chosen one, persuaded the Israeli Habadniks that the Messiah was none other than Schneerson himself. They built a house for him in Kfar Habad, a Tel Aviv suburb where most of them lived. Although Schneerson remained silent about being the Messiah, his followers in Israel became more and more persuaded that he would soon declare himself and move to Israel. Rabbi Adin Steinsaltz, a famous Talmudic scholar, declared that Schneerson was "the most likely person" then living who could turn out to be the long-awaited redeemer.

Other orthodox Jews in Israel were incensed by all this commotion. They denounced the Habad claims as blasphemy. According to *Time* (March 21, 1992), Eliezer Schach, a ninety-six-year-old Israeli rabbi who had long been Habad's chief enemy, said Schneerson was insane, an infidel, and a false Messiah. He even accused the Habadniks of eating pork and other nonkosher food.

Rabbi Schneerson was a short man with bright blue eyes and a large snow-white beard that made him look like a jolly Santa Claus. Born in the Ukraine and educated in Berlin and Paris as an engineer, he soon became the world leader of the Habad Lubavitchians. His International Educational Network, headquartered in Brooklyn's Crown Heights at 770 Eastern Parkway, is an outreach organization with its own TV cable network, and has a central aim of persuading Jews who have strayed from the faith to return to the fold. The outreach is said to take in some $100 million a year in contributions.

Rabbi Schneerson was much loved and admired by all New York City Jews even though conservative and reform Jews considered his messianic claims something of a joke. He was a magnetic preacher, and every Saturday thousands of admirers would jam into his Brooklyn synagogue to hear him. Every sermon was printed and faxed to his followers everywhere. A collection of his speeches runs to more than thirty volumes.

In 1993 the Schneerson mania in Israel and in New York City had not abated. A Brooklyn group called The International Campaign to Bring Moshiach [Messiah] took a full-page ad in the *New York Times* (Sunday, August 29). It featured a photograph of Schneerson surrounded in large print by the following message:

> The revered leader of world Jewry, Rabbi Menachem M. Schneerson, the Lubavicher Rebbe Shlita, issued a call that the time of our redemption has arrived, and Moshiach is on his way.
>
> The Rebbe, furthermore, stressed that he is saying this as a prophecy and asks all mankind to prepare themselves for the great day of redemption, with a personal commitment to increase in charity and good deeds.
>
> Over the years the Rebbe has inspired us, with his leadership, scholarship and prophesies time and time again. Now the Rebbe is telling us, as a prophecy, that Moshiach is on his way.
>
> Let us heed the Rebbe's call and let us all prepare for our own benefit, for that greatest of days, the ultimate purpose of G-d's creation.

I assume that the hyphen in "God" was intended to reflect the absence of vowels in Hebrew writing.

The Habad belief that Schneerson was the Messiah vanished after the rabbi died, following a stroke in 1994. He was ninety-two. Thousands of wailing mourners from all over the world attended his funeral at the Lubavitch international headquarters in Brooklyn.

ADDENDUM

At the close of my column I speculated that the hyphen in the word *God*, replacing the *o*, reflected the absence of vowels in ancient Hebrew. Many readers, some of whose letters appeared in *The Skeptical Inquirer* (July/August 2000), wrote to correct this error. Orthodox Jews hesitate to write "God" because it seems to be taking the name of the Lord in vain. Hence the hyphenated "G-d."

14

Oahspe

This essay first appeared in *The Encyclopedia of the Paranormal,* edited by Gordon Stein (Prometheus, 1996).

John Ballou Newbrough (1828–1891), named after John Ballou, a famous Universalist preacher, was born in a log cabin in Springfield, Ohio. His father was a Scot, his mother a Swiss. The mother was a devout spiritualist, and young John grew up sharing his mother's faith. He is said to have graduated from a Cincinnati medical college. I am not sure which one. Perhaps it was the city's Eclectic Medical College whose faculty stressed natural remedies in opposition to mainline medicine.

In 1849 young Newbrough joined the gold rush to California, where for several years he mined successfully. He became a strong champion of the civil rights of California's cruelly exploited Chinese laborers. Together with his friend John Turnbull, from Scotland, he prospected in the gold fields of Australia.

In 1859 Newbrough married Turnbull's sister Rachel. They settled at 128 West Forty-third Street, in Manhattan, where for twenty-three years Newbrough practiced dentistry. Always generous, much of his work for the poor was done free. Two boys and one girl came from the marriage. One boy died in infancy, the other was said to have graduated from

Columbia University as a civil engineer. The marriage degenerated, and in 1886 Newbrough sued for and won a divorce. A year later he married divorcée Frances Van de Water Sweet.

Newbrough was six feet, four inches tall and handsome, with a massive body and large hazel eyes. A great admirer of Battle Creek's Dr. Kellogg, he was a lifelong vegetarian and teetotaler. Following the doctor, he ate only one meal a day, avoided milk and eggs, and eventually abandoned all root vegetables because they grew without benefit of sunlight. This spare diet was said to have reduced his weight from 270 to 160 pounds.

Newbrough traveled widely in Europe and the Orient, lecturing to spiritualist groups while wearing brightly colored Oriental robes. His fame as a medium and automatic writer spread. Legends grew up around him. He was said to be able to paint pictures in total darkness, using both hands. It was claimed that he could close his eyes and read any book in any library. Without effort he could lift weights of a ton or more. His astral body was able to visit any spot on Earth. For two years he was active as a medium in "The Domain," a small spiritualist colony in Jamestown, New York.

One night at 4 A.M., as Newbrough himself tells it, he felt a hand on his shoulder and heard a voice urging him to wake. In his bedroom, flooded by a mysterious soft light, he saw the forms of beautiful, wingless angels. The voice told him he was destined for a special mission. He was to continue to abstain from flesh foods and to live a pure life, helping as many unfortunates as he could.

Ten years went by before he was awakened again by the mysterious light. The voice commanded him to buy a typewriter. Angels, it said, would control his fingers while he typed. These automatic typewriting sessions began on the predawn morning of January 1, 1881, and lasted until December 15 of the same year. Every morning before sunrise Newbrough pounded his Sholes typewriter, unaware, he insisted, of what he typed. All this, by the way, is from an account Newbrough gave in a long letter (dated January 21, 1883) that was published in the *Banner of Light*, Boston's leading spiritualist journal.

Many years before he began automatic typewriting, Newbrough had acquired the ability to write automatically in longhand. While sitting in spiritualist séances, "my hands could not lie on the table without flying into these 'tantrums.' Often they would write messages, left or right, forward or backward, nor could I control them any other way than by withdrawing them from the table."

After ten years of pure living and bathing twice a day, "a new condition of control came upon my hands." The control was through typing. The power descended upon him every morning before sunrise as he sat alone in his tiny apartment. One morning he glanced out the window and saw a long line of light that rested on his hands and "extended heavenward like a telegraph wire toward the sky. Over my head were three pairs of hands, fully materialized. Behind me stood another angel with her hands on my shoulders. . . . My looking did not disturb the scene; my hands kept right on printing, printing."

Was Newbrough lying? Did he actually experience these hallucinations? We shall probably never know.

For fifty weeks, Newbrough goes on in his letter, the angels controlled his typing for thirty minutes every morning before daylight. Suddenly the controls stopped. He was told by the angels to read for the first time what he had written, and to publish it as a book titled *Oahspe*. Newbrough printed the book himself in Boston, in 1882, on a press bought with money from seven anonymous associates. No author's name was on the title page. A revised edition was issued in 1891 and reprinted in London in 1910. The book was illustrated with pencil sketches drawn by the angels who controlled his hands. "A few of the drawings," he wrote, "such as Saturn, the Egyptian ceremonies, etc., I was told to copy from other books."

In 1960 Ray Palmer, a science-fiction writer and editor, published a photocopy of *Oahspe*'s first (1882) edition. From the 1891 edition Palmer reproduced twelve oil paintings by Newbrough of various *Oahspe* prophets, and photographs by Newbrough of ten cult children. In a second edition (1970) Palmer added *The Book of Discipline*, also taken from *Oahspe*'s second (1891) edition. Two years later, in another printing, Palmer supplied a seventy-page index. Palmer's final (1972) edition is called the "green *Oahspe*" because of its all-green hardcover.

Palmer is best known for being fired by the publisher of *Amazing Stories*, which Palmer edited, for hoaxing readers into believing that evil creatures called Deros live under the Earth's surface. As founder and editor of *Fate*, he was the first magazine publisher to promote the flying saucer craze, arguing that UFOs come from inside a hollow earth through a huge hole at the North Pole. See "Who Was Ray Palmer?" a chapter in my *New Age* (1991).

Oahspe teaches that there is one ultimate God who oversees a vast bureaucracy of lesser deities. This ultimate God has many names: I AM, Ormazda, Eloih, Creator, Most High, and Jehovih-Om. Jehovih is the Creator's masculine and positive aspect. Om is God's negative, feminine side. Like Christian Scientists and today's feminist theologians, Faithists like to think of the great I AM as Mother-Father, Him and Her. Jehovih is the term they most often use for the ultimate Creator. A lower God is assigned to each inhabited planet. Such gods are known as Emuts. The god assigned to Earth occupies a heavenly city called Hored. It is "situated over and above the mountains of Aotan in Ughoqui, to the eastward of UI."

Oahspe speaks of thousands of millions of gods, half of them are female. Closest to Jehovih are his countless Sons. Below the Sons are still lesser gods, and below the gods are billions of archangels, angels, and Lords. The Sons have such names as Sethantes, Ah'shong, Aph, Sue, Apollo, Thor, Osiris, I'hua Mazda, Yima, Lika, Uz, and Fragapatti. The goddesses have such names as Cpenta-Armij, Pathema. Harrwaiti, Dews, Cura, Yenne, Wettemaiti, and D'zoata.

Humans pop into existence at birth, with no previous lives, but after death they ascend from heaven to heaven in an endless series of adventures as they move toward perfection. Three days after death the souls of most mortals are carried by guardian angels, called ashars, to the lowest heaven where their souls are reborn in a "birth blanket." Angels who receive the souls are called asaphs.

> And the ashars shall make a record of every mortal, of the grade of his wisdom and good works; and when a mortal dies, and his spirit is delivered to the asaphs, the record shall be delivered with him; and the asaph, receiving, shall deliver such spirit, with the record into such place in these heavens as is adapted to his grade, where he shall be put to labor and to school, according to the place of the resurrections which I created.
>
> As ye shall thus become organic in heaven, with rulers, and teach-

> ers, and physicians; and with capitals, and cities, and provinces; and with hospitals, and nurseries, and schools, and factories, even so shall ye ultimately inspire man on the earth to the same things.
>
> And mortals that are raised up to dominion over mortals shall be called kings and emperors. As My Gods and My Lords are called My sons, so shall kings and emperors be called sons of God, through him shall they be raised up to their places, and given dominion unto My glory.

The heaven to which we go after death is in a higher dimension and therefore invisible. It occupies a region called Atmospheria because it lies within the atmosphere of the Earth's vortex. All heavenly bodies—suns, planets, moons—rose from and are sustained by rotating vortices of space-time.

Beyond Atmospheria are heavens of still higher dimensions in a vast region called Etheria. The food there is called heine, the drink haoma. One smells a potent perfume called homa. Etheria's dominant color is golden yellow. Everything in Etheria is made of ether, a solvent of corpor. Corpor is the matter of our corporeal world and our corporeal bodies. The unseen world is called Es in contrast to the visible world of Corpor. Inhabitants of Es are Es'eans. Those of Corpor are Corporeans. No two heavenly worlds are alike—"Every one differing from another, and with a glory matchless each in its own way." The heaven called Haraiti, and six others, were founded by Fragapatti.

There is no eternal hell. Those unprepared to enter the first heaven are imprisoned until they can be "weaned from evil." There is no final annihilation of the unredeemable.

Angels move rapidly about in nonmaterial Etherean spirit ships. The ships have such names as

> Arrow, Firre, Abattos, Adavasit, Airavagna, Airiata, Avalanza, Beyanfloat, Ballast Flags, Cowpon, Ese'lene, Koa'loo, Oniy'yah, Otevan, Ometer, Obegia, Piedmazar, Port-au-gon, Seraphin, Yista.

One ship, Koa'loo, is almost as large as the Earth. Some are propelled by musical vibrations, other by the vibrations of colors. Many Faithists today believe that these spaceships are responsible for the UFO sightings of recent decades.

Oahspe retells many of the tales in the Christian Bible, including the disobedience of Adam and Eve in eating the fruit of a forbidden tree. *Oahspe*'s account of Noah's Flood reflects the submergence of Whaga, a vast continent in the Pacific that later was called Pan. *Oahspe* is a word in the forgotten language of Pan. The "O" means earth, "ah" means air, and "spe" means spirit. The word is pronounced to imitate the sound of wind as it passes through trees, over oceans, and through mountains. The inhabitants of Pan were destroyed by Jehovih because of their wicked ways.

When Pan sank, the waters of the Pacific swallowed up the rich valleys of Mai, the wide plains of Og, the great capital of Penj, the temples of Khu, Bart, Gam, and Saing. Today's Zha'Pan (Japan) is a fragment of Pan that survived the sinking. After the submergence of Pan, human culture spread from Zha'Pan to Jaffeth (China). In *Oahspe* Asia is called Jud, Africa is Vohu, Europe is Dis, and America is Thouri.

Biblical personalities in *Oahspe* are given strange names. Satan is Anra'mainyus, Jesus is Joshu, Judas is Zoo-das, Adam is A'su, Abel is I'hin, Cain is Druk. Curiously, Eve remains Eve. A'su was made by the Creator out of se'mu. The Asu'ans, descendants of Asu, disobeyed Jehovih by eating the fruit of the Tree and were punished for it.

The admirable I'hins (descendants of Abel) were white and yellow, small and slender. The giant Druks (descendants of Cain) were brown and black, tall and stout—an evil race of murderers. Cohabitation between Druks and I'hins produced a hybrid race called I'huans. Cohabitation between Asu'ans and Druks, later between I'huans and Druks, produced the Yaks, or "ground people." They had long arms, curved backs, walked on all fours like apes, and were incapable of speech or of surviving death. The I'hins castrated them and made them slaves. The I'huans were copper colored and became the ancestors of the Ong'wee, or American Indians.

Moses and the Old Testament prophets, we are told, were early Faithists. They were called Eseans (Essenes). Joshu (Jesus) descended from this line. He was stoned to death at age thirty-six by Jews who worshiped heathen gods. Forty years later a phony deity named

Looeamong called himself Christ. His warrior tribes became the early Christians. The Christians, says *Oahspe*, are "warriors to this day."

Rather than reading all of *Oahspe*, which I find too boring to keep my eyes open, I let the plot summary in the *Encyclopedia of Religion and Ethics* finish the story:

> Looeamong, with the other Triunes, Ennochissa and Kabalactes, endeavoured to overthrow Jehovih, assuming the names of Brahma, Budha, and Kriste to combat Ka'yu (Confucius), Sakaya (Buddha), and Joshu. But Looeamong failed to keep his word to his chief angel warrior, Thoth, or Gabriel, who rebelled in consequence, and raised up Muhammad. Muhammadism is to perish first, then Brahmanism, then Buddhism, and finally Christianity. During the period treated by the Book of Es (c. 1448–c. 1848) there is an abrogation of revelations, ceremonies, etc., and liberty of thought begins to prevail. Melkazad is divinely sent to inspire a migration to Guatama, and he raises up Columbo to discover it to broaden the sphere of Jehovih's kingdom and to aid in overthrowing the Triunes and Thoth. Then Looeamong inspires his followers (Roman Catholics) to punish heresy, thus giving rise to Protestantism, which also is inspired by evil spirits. The Pilgrim Fathers were inspired by the God, but corrupted by Looeamong; the Quakers were Faithists at heart. Thomas Paine was inspired by Jehovih, the other chief men raised up by God, to establish the foundation of Jehovih's kingdom with mortals, being Jefferson, Adams, Franklin, Carroll, Hancock, and Washington. During the decay of Looeamong's kingdom petty Drujan Gods set up little principalities, such as Methodists, Presbyterians, and Baptists, while Pirad founded the Mormons, Lowgannus the Shakers, and Sayawan the Swedenborgians.

For *Oahspe*'s high praise of Thomas Paine's attack on Christian doctrines, at the same time defending theism and a hope for immortality, see chapter 13 of the Book of Es. Chapter 20 of the same book tells how angels came to Lincoln in dreams and inspired him to free the slaves.

Oahspe speaks of stars, planets, and moons as condensing from huge vortices or swirls in the "etherian firmament." Our Sun is at the center of a mammoth vortex. Its planets and moons are the centers of smaller vortices. By stretching things a bit, one could take this to be prophetic of general relativity in which massive objects like stars and planets bend space-time into what could be called surrounding vortices that generate gravity.

Jehovih states that our solar system travels an orbit so huge that it takes 4,700,000 years to complete one revolution. Astronomers today estimate that the solar system completes its orbit around the center of the Milky Way galaxy in about 200,000,000 years. However, in Newbrough's day, 4,700,000 was not such a bad guess.

When Ray Palmer's *Mystic Magazine* reprinted excerpts from *Oahspe* (February 1955), in an article headed "The Most Amazing Book in the World," an unidentified writer (perhaps Palmer) supplied footnotes.

Oahspe clearly states that it is not to be taken as containing absolute truth. The following lines appear in *Oahspe's* first paper: "Not infallible is this book, OAHSPE; but to teach mortals how to attain to hear the Creator's voice and to see His heavens, in full consciousness, whilst still living on the earth; and to know of a truth the place and condition awaiting them after death."

Oahspe is written in the style of the King James Bible, with lots of "thees," "yeas," "beholds," and other biblical words and phrases. Like the Bible it is divided into books, in turn divided into chapters, in turn divided into numbered verses. Above all, *Oahspe* contains thousands of strange proper names for persons and places, and neologisms never used before or since.

Faithists delight in claiming that all sorts of scientific truths in their sacred book, unknown at the time, have been confirmed by recent discoveries. An Owl Press advertisement for *Oahspe* declares that the book's science "is today being confirmed by space-satellites and new archaeological discoveries of ancient races, dead cities and civilizations."

Oahspe divides all living humans into two classes. Those who accept the new revelation are the Faithists. Outsiders who are not Faithists are called Uzians.

Uzian reviewers of *Oahspe* were merciless in their criticism, although they marveled that any one man could single-handedly produce such a monumental work. If *Oahspe* is a true revelation, said the critics, it has to be the greatest book ever written, even more important to mankind than the Bible. If not true, it is either the work of a psychotic or a monstrous hoax.

Soon after *Oahspe* was published, plans for a colony of Faithists began to jell. A small commune was set up in Woodside, New Jersey, later moving to Pearl River, New York. It did not last long.

Enter Andrew M. Howland, a wealthy Quaker and businessman from Massachusetts. He was Faithism's most notable convert, and almost as mad as Newbrough. The two became great friends.

A book in *Oahspe* titled "The Book of Shalam" recommended the establishment of a colony to take care of unwanted orphans. In the mid-1880s, under the direction of angels, Newbrough and Howland founded the Children's Land of Shalam, near the village of Dona Ana in southern New Mexico, on the east banks of the Rio Grande. When the colony failed, Howland and other backers lost more than a million dollars.

The first settlers in Shalam came mainly from the Pearl River colony—only about twenty at first, but the ranks soon swelled to fifty. These historical details, I must add, as well as those to come, are based entirely on a lengthy article titled "The Land of Shalam," by a Mrs. K. D. Stoes. It appeared in the *New Mexico Historical Review* 33, nos. 1 and 2 (January and April 1958). This history was reprinted as a sixty-eight-page booklet, currently available from the Universal Brotherhood of Faithists, in Tiger, Georgia. It is the only history of Shalam known to me. I have no idea how accurate it is.

For five years, Mrs. Stoes writes, people came to Shalam and left. They included "adventurers, religious fanatics of dubious faiths, habitual new creeders, and a few mentally deficient." Accommodations were available for a hundred. Money flowed in from Eastern philanthropists, most of whom knew nothing about Newbrough or the strange doctrines of *Oahspe*. Unwanted foundlings from large cities were taken in. There were no racial bars. The orphans were white, black, brown, and yellow—babies the world did not want. They were given Oahaspian names such as Pathocides, Astrafm Thouri, Hiatisi, Hayah, Thalo, Ninya, Havalro, Hiayata, Des, Fiatsi, Voh nu, Whaga, and Ashtaroth. No records of parents were kept. In later years, many of the children, grown to adults, would desperately try to learn who their parents were. With the help of cheap Mexican labor, buildings were erected. Howland became known as Father Tae, and the first building to go up was designated the Temple of Tae.

Mrs. Newbrough was known to all as "Mama." Everyone in the colony followed Dr. Kellogg's recommendations of a strict vegetarian diet and no more than two meals a day. By the late 1880s the colony held about twenty-five Faithists and some fifty children.

Great emphasis was placed on the properties of colors. Yellow was the most sacred. Blue was a cold color that could cure baldness and induce sleep. Gastritis was relieved by water from blue bottles after they had been radiated with sunlight. Pictures of the great prophets mentioned in *Oahspe* were painted by Newbrough on the walls of a building called the Faturnum. He is said to have painted them with both hands at once. Angels constantly gave him advice on how to manage his colony.

In 1891, at the age of sixty-three, Newbrough died of pneumonia along with other Faithists in an influenza epidemic that swept the region. Howlands was in Boston at the time, seeing to the printing of a revised second edition of *Oahspe*. Newbrough died before it left the press. A lifelong Mason, his body eventually found rest in a Masonic cemetery in Las Cruces.

On his tombstone, under the Faithist logo of cross and leaf, are the words: "Unto Thee, Jehovih-Creator, be praise and thanks for brother John B. Newbrough, June 5, 1828–April 23, 1891, through whose hands *OAHSPE*, the New Bible, was transcribed for the World."

Howland, now the colony's patriarch, with his large gray beard, blue eyes, long hair, white trousers, and sandaled feet, took over Shalam's management.

In 1893 he married Newbrough's widow. For a few years the colony prospered. Howland established a chicken farm. Eight windmills pumped water to the buildings from the Rio Grande. There were a machine shop, general store, stables, stock pens, and a bee aviary to provide honey. Efforts were made to build Levitica, a town of twenty small houses. Residents were brought in from Kansas City and elsewhere. They began to quarrel. Some fancied that the colony permitted free love. Disgusted, Howland packed them off to where they came from.

Like so many other small religious colonies with weird doctrines, Shalam slowly disintegrated. Funds dwindled. Storms and floods destroyed property. By 1901 the commune was bankrupt. About twenty-five children were placed in foster homes or sent to orphanages in Denver and Dallas. Booker T. Washington adopted a bright little black boy called Thail. The Howlands blamed the failure of Shalam on evil angels. For a couple of years the couple tried to preserve the faith in California. They and a few followers settled in El Paso where Howland died in 1917, at age eighty-three. Five years later his wife followed him upward through the stars.

In 1918 the American poet Ella Wheeler Wilcox published an autobiography titled *The World and I*. Mrs. Wilcox is now forgotten by both critics and the general public, but she was then at the height of her fame. In addition to writing poetry, novels, and essays, she was a leader of what was then called New Thought, a precursor of today's New Age. Ella was the Shirley MacLaine of her day—theosophist, Spiritualist, and true believer in all things psychic and occult. After her husband died in 1916, she began a desperate search around the nation for spiritual solace and for a genuine communication from her husband's spirit. Eventually she established contact with him by way of a Ouija board, but before this her search took her to California where she visited the remnants of the Oahaspian cult. She found them to be "a strange and earnest handful of men and women, following altruistic ideals, but leaving me sadder than before I visited them. They seemed to have eliminated from life on earth all idea of beauty."

References

Gardner, Martin. "Oahspe." Chapter 9 in *Urantia: The Great Cult Mystery*. Amherst, N.Y.: Prometheus Books, 1995.

Gray, Louis G. "Oahspe." In *The Encyclopedia of Religion and Ethics*, edited by James Hastings. New York: Scribner's, 1908.

Newbrough, John B. *Oahspe*. Boston: Privately printed, 1882. Revised edition 1891. Reprinted by Ray Palmer in enlarged editions of 1960, 1970, and 1972.

Historical Review 33, nos. 1 and 2 (January and April 1958). Reprinted as a booklet by the Universal Brotherhood of Faithists (UB), n.d.

Wiley, Elnora W. *Inside the Shalam Colony*. Los Alamos, N.M.: Document Shop, 1991.

15

The Vagueness of Krishnamurti

This essay first appeared in *The Skeptical Inquirer* (July/August 2000). In it I will sketch David Bohm's sad life and his strange relationship with the Indian guru Jiddu Krishnamurti.

David Bohm was born in Wilkes-Barre, Pennsylvania, in 1917. When he obtained his doctorate in physics under J. Robert Oppenheimer, at the University of California, Berkeley, Bohm was a dedicated Marxist and a strong admirer of Lenin, Stalin, and the Soviet system. These opinions drew the fire of Senator Joseph McCarthy. Bohm's refusal to name names resulted in his indictment for contempt of Congress. Princeton University, which had hired him, let him go. No other university in America wanted him. After brief periods of teaching in Brazil, Israel, and England, he finally became a professor at London's Birkbeck College where he remained until he retired.

As Bohm grew older, he became increasingly preoccupied with Eastern mysticism and parapsychology. The Indian philosopher Krishnamurti became a good friend. The "All is One" aspect of Buddhism and Hinduism, and the pantheism of Hegel and Alfred North Whitehead, strongly influenced Bohm's view of the universe. He became convinced that being is multidimensional, with infinite levels in both directions—levels far beyond our comprehension. On Newton's level the universe is deterministic and mind independent. On the quantum level it rests on uncertainty and chance, with tinges of solipsism. Below the quantum level, Bohm believed, is a subquantum world in which determinism and reality return. And below that? The levels are endless. Ultimate truths are forever beyond our grasp. I do not know whether Bohm believed in reincarnation or personified the Unknowable as the Hindu god Brahman, the ultimate ground of being about whom nothing can be said.

"If Bohm's physics, or one similar to it," Gary Zukav writes in his popular New Age book *The Dancing Wu Li Masters* (1979), "should become the main thrust of physics in the future, the dances of East and West could blend in exquisite harmony. Do not be surprised if physics curricula of the twenty-first century include classes in meditation."

For another typical example of how occult journalists have latched onto Bohm, see Michael Talbot's *The Holographic Universe* (1991). Talbot buys just about everything on the paranormal landscape including palmistry, UFOs, poltergeists, and dermo-optical perception—the ability to see with fingers, nose, and armpits. His book's main theme is that Bohm's quantum potential field accounts for all paranormal wonders. Curiously, Talbot doesn't mention astrology, even though Bohm's quantum potential might offer a good basis for it.

Bohm's disenchantment with Soviet Communism did not come until 1956 when Nikita Khrushchev delivered his blistering attack on Stalin, and the magnitude of Stalin's purges and the slave labor camps became widely known. It was a crushing blow that plunged Bohm into the second of his periodic depressions. (The first began when he was twenty-six, and lasted two years.) For six months Bohm underwent intensive Freudian psychoanalysis before this second depression lifted.

Ex-Communists and fellow travelers have a habit of turning from Marxism to another ideology, often Catholicism or some other religion. In Bohm's case it was a bounce toward Buddhism and Hinduism, and the teachings of Krishnamurti. After decades of close friendship, with unbounded admiration largely on Bohm's side, the two had a bitter falling out. Krishnamurti always had a low opinion of physics, and Bohm's pilot wave theory in particular. He had a cruel way of treating Bohm as if he were a stupid child unable to fully appreciate his (Krishnamurti's) vast wisdom.

Krishnamurti (1895–1986) was one of the most peculiar gurus ever to come out of Mother India. In 1908, this thin, frail, shy lad, of Brahmin birth, was discovered by Annie Besant and Charles Leadbeater, the most famous disciples of Madame Blavatsky. They became convinced that young Jiddu was the new messiah, or world teacher, and the incarnation of Lord Bodhisattva Maitreya, the fifth Buddha. Leadbeater, who claimed to be clairvoyant, saw all this when he viewed Jiddu's aura. Besant adopted him as her son and raised him as a theosophist. In 1910, Krishnamurti's first book, *At the Feet of the Master*, was published by England's Theosophical Society. It was said to have been written by Krishnamurti who used the pseudonym of Alcyone, when he was fifteen.

A young Krishnamurti with Annie Besant (Corbis)

In 1911, Besant, then the international president of the Theosophical Society, founded the Order of the Star of the East. Krishnamurti was the rising Star. In 1922 Annie purchased six acres in Ojai, California, where Krishnamurti eventually settled, and which became the headquarters of the still-flourishing Krishnamurti Foundation.

Jiddu's father lost a lawsuit trying to regain custody of his son. His lawsuit accused Leadbeater, who was probably gay, of having had sexual relations with Jiddu.

In 1922 Krishnamurti had a spiritual awakening which *Harper's Encyclopedia of Mystical and Paranormal Experience*, edited by Rosemary Ellen Gulley, describes as follows:

> He suffered excruciating headaches, visions, and convulsions, shuddering and moaning, and semiconsciousness, much as a person possessed. These seizures and spiritual manifestations lasted for several years and formed the basis for Krishnamurti's later orientation. He called the ordeal "an inward cleansing."

Krishnamurti tried to enter Oxford, but failed its entrance examination. He never got a college degree. In 1929 he made a cleansing break with his theosophical upbringing by disbanding the Order of the Star. A year later, to Annie Besant's sorrow, he resigned from the Theosophical Society. Henceforth he would travel around the globe, giving talks and conducting dialogues in which he taught a vague form of consciousness raising unrelated to any religion, and based on his own techniques of meditation and self-improvement.

As Bohm's friend and collaborator David Peat tells it in his biography of Bohm, *Infinite Potential* (1997), young Krishnamurti actually believed for a time that he was indeed the incarnation of Lord Maitreya, and the true successor to Jesus. His consciousness and that of Maitreya had merged; the "beloved" spoke through him. Although Krishnamurti outgrew the theosophical nonsense Besant and Leadbeater had drummed into him, he never stopped believing that he and he alone among living mortals knew the truth about everything. His teaching was a mix of dull platitudes and murky phrases such as "the observer is the observed," "thinking is the thought," "choiceless awareness," and that to be transformed one must "die to the moment." He was convinced that when a person was radically changed through proper meditation there were actual mutations in the brain!

Krishnamurti's name was on more than forty books, as well as on endless audio and video tapes. *The Ending of Time* (1985) was a book coauthored with Bohm. I have done my best to try to read some of these books without falling asleep. It is hard to understand how the author of such vapid ideas could have mesmerized listeners, most of them women, when he lectured, and to have captured the admiration of a great physicist. His lines are like those in Lewis Carroll's "Jabberwocky." As Alice remarked, they seem to mean something, but it's hard to pin down just what. There is never a hint in Krishnamurti's writings of a personal God or the survival of personality after death. He almost never refers to or quotes from any other thinker. His vision is a kind of watered-down Buddhism in which the key message is that everything is interconnected, and one must live in the moment, without fear, and accept everything that happens with resignation and tranquility. The same infuriating vagueness permeates books written by his admirers.

To give you a glimpse into Krishnamurti's vagueness, here are a few typical excerpts from his talks:

> . . . There is no such thing as doing right or wrong when there is freedom. You *are* free and from that centre you act. And hence there is no fear, and a mind that has no fear is capable of great love. And when there is love it can do what it will.
>
> Death is a renewal, a mutation, in which thought does not function at all because thought is old. When there is death there is something totally new. Freedom from the known is death, and then you are living.
>
> When you love, is there an observer? There is an observer only when love is desire and pleasure. When desire and pleasure are not associated with love, then love is intense. It is, like beauty, something totally new every day. As I have said, it has no yesterday and no tomorrow.
>
> As long as there is a time interval between the observer and the observed it creates friction and therefore there is a waste of energy. That energy is gathered to its highest point when the observer is the observed, in which there is no time interval at all. Then there will be energy without motive and it will find its own channel of action because then the "I" does not exist.
>
> To see what you actually are without any comparison gives you tremendous energy to look. When you can look at yourself without

comparison you are beyond comparison, which does not mean that the mind is stagnant with contentment.

You can face a fact only in the present and if you never allow it to be present because you are always escaping from it, you can never face it, and because we have cultivated a whole network of escapes we are caught in the habit of escape.

We might be able to modify ourselves slightly, live a little more quietly with a little more affection, but in itself it will not give total perception. But I must know how to analyse which means that in the process of analysis my mind becomes extraordinarily sharp, and it is that quality of sharpness, of attention, of seriousness, which will give total perception. One hasn't the eyes to see the whole thing at a glance; this clarity of the eye is possible only if one can see the details, *then jump*.

It was not just Bohm who fell under the sway of Krishnamurti's charisma. He strongly influenced such writers as Joseph Campbell, the poet Robinson Jeffers, Henry Miller, Aldous Huxley, and Alan Watts who churned out popular books about Zen Buddhism. George Bernard Shaw once called young Krishnamurti "the most beautiful human being" he ever saw. After visiting Krishnamurti's castle in Holland, Campbell wrote in a letter: "I can scarcely think of anything but the wisdom and beauty of my friend." In another letter he said, "Every time I talk with Krishna, something new amazes me."

There were two Krishnamurtis. One was the persona presented to the world through lectures and books; a man without ego who led a sanctified life of celibacy and high moral purity. The other Krishnamurti was a shadowy, self-centered, vain man, capable of sudden angers and enormous cruelty to friends. He was also a habitual liar. Krishna, as his friends called him, freely admitted his compulsive lying. He blamed it on simple fear of having his deceptions detected.

Krishna's closest associate was Raja, whose full Indian name was Desikacharya Rajagopalacharya. A native of India, Raja was as handsome as Krishna, and for almost thirty years a devoted disciple who served as his master's business manager, secretary, literary agent, and editor. It was Krishna's good friend Aldous Huxley who introduced Raja to his editor at Harper and Row, the firm that published Krishna's many books. Krishna had little interest in writing or publishing, but he allowed Raja to cobble books out of his talks and notebooks, and to edit this material into volumes.

Raja's wife, Rosalind, was a beautiful American Caucasian who grew up in Hollywood, a friend of movie stars, who almost became a professional tennis player. Toward the end of Krishna's life an astonishing revelation came to light. For nearly thirty years, unknown to Raja, Rosalind had been Krishna's mistress! As such, she had undergone a miscarriage and several abortions, all miraculously kept secret from her husband.

Raja forgave his wife and never ceased loving her, but a rift between Raja and Krishna grew steadily wider until finally they became bitter enemies. Twice Krishna unsuccessfully sued Raja for mishandling funds, and Raja in turn sued Krishna for slander. All three lawsuits were finally settled out of court. The two former friends never reconciled.

Rosalind's passion for Krishna cooled when she discovered he was having another secret affair, this time with a shy, beautiful young woman, Nandini Mehta. Raja's passion for Rosalind also dimmed when he fell in love with Annalisa Beghé, a Swiss-Italian who was twenty-three years his junior. After he and Rosalind finally obtained a Mexican divorce, Raja and Annalisa married.

You can learn all the sordid details about these surprising events in a splendid biography of Krishna, *Lives in the Shadow with J. Krishnamurti* (1991), by Radha Rajagopal Sloss, the daughter of Rosalind and Raja. Now married to mathematician James Sloss, a professor at the University of California, Santa Barbara, Radha is also the author of *India Beyond the Mirror* (1988).

A story in Mrs. Sloss's book reveals how little Krishnamurti understood science. America's funniest and nuttiest medical quack was Dr. Albert Abrams. You can read about him in my *Fads and Fallacies*. Abrams invented a bizarre electrical machine which he claimed could diagnose all ailments from a drop of blood, and could cure terrible diseases by electrical rays. Like the writer and socialist Upton Sinclair, Krishna became convinced that Abrams's machine did everything he claimed it did. He sent Abrams a blood sample, and was told he had cancer in his intestines and left lung, and syphilis in his spine and nose. After being treated for a while by Abrams's machine, Krishna said he felt much better.

Radha likens her father to the Hindu god Vishnu, the preserver, and Krishna to Shiva, the destroyer. "The division between Krishnamurti himself will cast a very dark shadow on all he has said or written," Radha concludes. "Because the first thing the readers will say, is: 'If he cannot live it, who can?' "

After learning about Krishnamurti's secret love affair with his best friend's wife, Bohm felt betrayed. Perhaps this plunged him into his third and final deep depression. Hospitalized, suffering from paranoia and thoughts of suicide, Bohm underwent fourteen episodes of shock therapy before he recovered sufficiently to leave the mental hospital. Earlier triple bypass surgery on his heart had been successful, but his death in 1991, at age seventy-five, was from a massive heart attack. Krishnamurti had died six years earlier, at his home in Ojai, of pancreatic cancer. His body was cremated.

Bohm's creative work in physics is undisputable, but in other fields he was almost as gullible as Conan Doyle. He was favorably impressed by Count Alfred Korzybski's *Science and Sanity*, with the morphogenic fields of Rupert Sheldrake, the orgone energy of Wilhelm Reich, and the marvels of parapsychology.[1] For a while he took seriously Uri Geller's ability to bend keys and spoons, to move compasses, and produce clicks in a Geiger counter, all with his mind.

Bohm also flirted with panpsychism, the belief that all matter is in some sense alive with low levels of consciousness. "Even the electron is informed with a certain level of mind," Bohm said in an interview published in *Quantum Implications: Essays in Honor of David Bohm* (1987), edited by Basil Hiley and David Peat. Bohm's later writings swarm with

[1] Sheldrake's latest book, published by Crown in 1999, is titled *Dogs That Know When Their Owners Are Coming Home And Other Unexplained Powers of Animals*. The other animals with psychic powers include cats, parrots, horses, monkeys, sheep, pigs, rabbits, and even chickens! On Sheldrake's earlier bizarre claims, see Chapter 15 in my 1988 collection of *Skeptical Inquirer* columns, *The New Age: Notes of a Fringe Watcher*.

neologisms such as *holomovement, rheomode, levate, enfoldment, somasignificant*, and *implicate and explicate levels of reality*.

In his biography of Bohm, David Peat tells how Bohm carried with him a key bent by Uri Geller as if it were a holy relic. When the key later disappeared, Bohm took this to be Geller's psychokinetic powers at work from a distance. When the key was found an hour later, he believed this to be another paranormal event! Bohm's close associate Basil Hiley at once recognized Geller as a charlatan. He often warned Bohm that if he appeared to endorse Geller it would damage their work. Bohm agreed to back away from Geller. As Hiley said to Peat, Bohm often had to be saved from idiots.

Bohm's Eastern metaphysics, even though it helped shape his interpretation of quantum mechanics, should not be held against the potential fruitfulness of his pilot wave theory. In a similar fashion Isaac Newton's Biblical fundamentalism and his alchemical research cast no shadows over his contributions to physics. Nor did Kepler's belief in astrology throw doubts on his great discoveries.

Einstein once said that his misgivings about the Copenhagen interpretation of QM came from something he felt in his little finger. For decades his intuitions were ridiculed by Niels Bohr and his followers. They never ceased to express sorrow over how the great man had deserted them by refusing, in his senior years, to view QM as a beautiful, complete theory, in no need of being replaced or modified.

It is too bad that Einstein did not live to see an increasing number of top physicists, such as Roger Penrose, Jeffrey Bub, and John Bell, who suspected, and today suspect, that the old maestro may turn out to be right after all.

ADDENDUM

Four letters from admirers of Krishnamurti appeared in *The Skeptical Inquirer* (November/December 2000) accusing me of undue harsh treatment of Jiddu. For a less harsh recent biography, see *Star in the East: Krishnamurti, the Invention of a Messiah*, by Roland Vernon, published in England by Constable in 2000.

Will Stevens, reviewing this book in the British magazine *The Skeptic* (I failed to date my clipping), writes that Jiddu comes through as "egotistical and infantile," whose writings "appear to be little more than windy pantheistic rhetoric." Stevens concludes that Vernon's book is a "better biography than its subject deserved."

My earlier *Skeptical Inquirer* column on David Bohm's heretical "Guided Wave Theory" of Quantum Mechanics appeared in the May/June 2000 issue. The column is reprinted in my collection *Did Adam and Eve Have Navels?* (Norton, 2000).

PART IV

Literature

16

Chesterton's *The Man Who Was Thursday*

This chapter first appeared as an introduction to *The Man Who Was Thursday* by Gilbert Chesterton (Ignatius, 1999).

The Man Who Was Thursday, a masterpiece by G. K. Chesterton, revolves around two of the deepest of all theological mysteries: the freedom of the will and the existence of massive, irrational evil. The two mysteries are closely related.

In Chesterton's comic fantasy, which he calls on the title page "A Nightmare," free will is symbolized by anarchism. Man's freedom to do wicked things, as Augustine and so many other theologians of all faiths have said, is the price we pay for freedom. If our behavior were entirely determined by how our brain is wired by heredity and environment, then we would be mere automatons with no more genuine free will or self-awareness—two names for the same thing—than a vacuum cleaner. But we are not automatons. We have a knowledge of good and evil and a freedom to choose, within limits, of course, between the two. Somehow our choices are not totally determined, yet somehow they also are not random, as if decisions were made by shaking tiny dice inside our skull. This is the dark, impenetrable paradox of will and consciousness. "I see everything," Gabriel Syme shouts in the book's last chapter. "Why does each thing on the earth war against each other thing? . . . So that each thing that obeys law may have the glory and isolation of the anarchist."

The anarchist movement of Chesterton's time, with its fanatical bomb tossers, has happily faded, but individual anarchists are still with us. A Timothy McVeigh blows up a federal building because he hates the federal government. A Ted Kaczynski blows up strangers because he hates modern technology. Islamic extremists blow up buildings and airplanes because they hate Israel and the United States. Irish Catholics and Protestants explode bombs because they hate each other. Such are some of the horrors we pay for the mysterious gift of free will.

Henry James, the father of William, said it eloquently in a letter quoted by Ralph Barton Perry in the first volume of his *Thought and Character of William James* (1935, p. 158):

> Think of a spiritual existence so wan, so colourless, so miserably dreary and lifeless as this; an existence presided over by a sentimental deity, a deity so narrow-hearted, so brittle-brained, and putty-fingered as to be unable to make godlike men with hands and feet to do their own work and go their own errands, and content himself, therefore, with making spiritual animals with no functions but those of deglutition, digestation, assimilation. . . . These creatures could have no *life*. At the very most they would barely *exist*. Life means individuality or character; and individuality or character can never be *conferred*, can never be *communicated* by one to another, but must be inwardly wrought out by the diligent and painful subjugation of evil to good in the sphere of one's proper activity. If God made spiritual sacks, merely, which he might fill out with his own breath to all eternity, why then of course evil might have been left out of the creature's experience. But he abhors sacks, and loves only men, made in his own image of heart, head and hand.

The 168 persons murdered by McVeigh's fertilizer bomb were just as irrationally killed as if an earthquake had leveled the building. And this takes us to the other deep mystery of Chesterton's nightmare, the mystery of natural evil. Of course, this is no mystery for an atheist. It's just the way the world is. But for a theist of any faith it is the most terrifying of all riddles. How can an all-powerful, benevolent God permit so much needless pain? As Gögol asks Sunday, like a small child questioning his mother, "I wish I knew why I was hurt so much."

The reality of such vast amounts of suffering provides atheists with their most powerful argument. Earthquakes, made inevitable by stresses in strata below the earth's surface, can snuff out the lives of thousands. Little children die of cancer. Millions can be killed by epidemics such as the Black Death of the fourteenth century, a bubonic plague that wiped out almost half of Europe.

The only possible way a theist can escape from the atheist's charge—either God is malevolent or there is no God—is to view Nature as the back of reality. Beyond what Lord Dunsany liked to call "the fields we know" there is a larger, wholly other unseen realm. Logic cannot prove its existence, and science is helpless in efforts to penetrate it, but by a leap of faith we can escape despair by looking forward to a life beyond the grave where God will in some manner, utterly beyond our understanding, rectify the mad injustices of the fields we know. This is the great hope that glows at the heart of theism and at the core of Chesterton's melodrama.

Many readers over the decades have found it difficult to understand who Sunday is. In the first chapter of F. Scott Fitzgerald's *This Side of Paradise*, the protagonist is said to have liked *The Man Who Was Thursday* but without understanding it. An unsigned reviewer in the *Aberdeen Free Press* (March 12, 1908) ended his review by saying he was entertained by G. K.'s "brilliant prose" but put the book down "with no earthly idea" of what it was all about.

Who, then, is Sunday? Chesterton himself made it plain enough not only in his novel . . . but also in comments about the novel that I have gathered in G. K.'s book's appendix. Sunday is simply Nature, or the Universe when seen as distinct from the Creator. The God of Judaism, Christianity, and Islam has two aspects that theologians like to call transcendence and immanence. God is totally beyond the universe and our comprehension yet at the same time closer to us than breathing, as the Bible says, or, in the Koran's words, closer than the main artery in our neck. Sunday is God's immanence. He is Nature, the Universe, with its unalterable God-given, God-upheld laws that seem so obviously indifferent about our welfare.

Sunday, like Nature, has a front and back side. From the back he resembles what Chesterton calls in *The Uses of Diversity* (chapter 9) a "semi-supernatural monster." From the front he looks like an angel. Nature lavishes on us a thousand gifts that make us happy and grateful to be alive, yet the same Nature can destroy entire cities with seemingly random earthquakes. It can drown us with floods, kill us with tornadoes and diseases. Ultimately it will execute us.

Atheist and theist alike must face the fact that Nature cares not a rap whether you or I live or die, or even whether the human race will survive. There is no guarantee that some day a giant comet or asteroid will not strike the earth and obliterate all life. We may destroy ourselves with nuclear war. There is no assurance that man will not ultimately vanish like the dinosaurs.

Throughout Chesterton's nightmare are numerous hints that Sunday is pagan Nature. He is monstrously huge and shapeless. When he stands he seems to fill the sky. His room and clothes are neat, but he is absent-minded and at times his great eyes suddenly go blind. Did G. K. make his eyes blue because that is the color of the sky? Sunday's white hair suggests his great age. We are told he never sleeps. Like God's omnipresence, he can be in six places at once. He is capable of smashing a person "like a fly." He resembles a human but actually "is not a man." Like Pan he is half human, half animal.

"We are not much," Ratcliffe says, "in Sunday's universe." Those who come in contact with Sunday fear him the way they fear the "finger of God." Who can contemplate the universe—billions of flaming stars in every galaxy and billions of galaxies—without being profoundly disturbed by a sense of wonder combined with sheer terror. This is a sampling of how the six men, Monday through Saturday, react to Sunday. He arouses in them that strange mixture of awe and fear which Karl Otto, in *The Idea of the Holy*, calls the *mysterium tremendum*.

Chesterton liked to imagine that God has a sense of humor. Sunday is described as a "Thing" capable of shaking with laughter "like a loathsome and living jelly." Nature has its wildly comic side. It enjoys playing "good-natured tricks" "so big and subtle" that we could never have thought of them until we see them—jokes like the pelican, the hornbill, the elephant.

Einstein once said, in an often quoted remark, that God (by which he meant Nature, or the "God" of Spinoza) is subtle but not malicious. It is not generally known that later in life, after the development of quantum mechanics, Einstein admitted in a letter that perhaps he was mistaken. God may be malicious after all.

Einstein was not thinking of such malicious actions as earthquakes and pestilences—Spinoza's God is indifferent about such things—but of the many subtle paradoxes of quantum theory. Consider the notorious EPR paradox, the letters standing for Einstein and two colleagues who first discovered it. A pair of particles is produced by an event that sends them off in opposite directions. Their production requires that they have opposite spins. In quantum theory spins have no direction until they are measured. Yet no matter how far apart the particles go, perhaps light years from each other, they remain "entangled" in such a way that when the spin of one is measured, the wave function of the two-particle system is said to "collapse," and the other particle instantly acquires a spin opposite to that of the particle measured.

Einstein called this "spooky action at a distance." The paradox is not resolved by saying that the particles are always part of the same quantum system with a single wave function that collapses when one particle is measured. The mystery is how the particles manage to stay entangled, or "correlated," when relativity theory makes it impossible for information to travel faster than light. For Einstein, in his later years, the EPR paradox was one of the "Old One's" malicious little jokes.

Nature swarms with a thousand other good-natured pranks that remain unexplained. When scientists ask questions about them, they often get answers that appear to be nonsense. At the moment astronomers are mystified by evidence that seems to imply that the universe is younger than some of its stars. In the Old Testament's Book of Job, of which Chesterton was especially fond, Job does his best to force God to explain why he, Job, a good man, must suffer such agonies. God answers Job's riddles by flinging other riddles at him. Who do you think you are, God says, to question the wisdom and intentions of your creator? Where were you when I made the universe? "The *Iliad* is only great," wrote G. K. in an essay on nonsense (*The Defendant*, 1907), "because all life is a battle, the *Odyssey* because all life is a journey, *The Book of Job* because all life is a riddle."

In 1907, the year before Thursday was published, Chesterton wrote an introduction to the Book of Job. (It is reprinted in *GK as MC*, 1929.) Garry Wills, in his introduction to *The Man Who Was Thursday*, calls this Chesterton's "most important essay, written on the book that most profoundly influenced him all his life." This essay, writes Wills, "could almost stand as a commentary on the novel."

> The "Council" and the "Accuser" are, in the last scene, direct references to the Book of Job. The final chase through monstrous scenes, thronged with trumpeting and incredible beasts, is a glimpse of that animal world which Jehovah called up for Job. Syme is answered by the elephant, as Job was by Behemoth. These echoes multiply in the final chapter as the Sons of God shout for joy in the strange dance the Council witnesses. The parallels are finally established by Bull's quotation: "Now there was a day when the sons of God came to present themselves before the Lord, and Satan came also among them."

God's evasive, irrelevant replies to Job are parodied by the nonsense messages Sunday flings at his pursuers while he is being chased across London. Nature is forever confronting scientists with phenomena they cannot fathom. No one can catch Sunday. No one can discover the ultimate reasons for why the universe exists or why it is structured the way it is.

"What am I?" Sunday roars in chapter 13. (Note: he says "what" not "who.") He goes on to add that science will never discover everything. "You will understand the sea, and I shall be still a riddle; you shall know what the stars are, and not know what I am. Since the beginning of the world all men have hunted me. . . . But I have never been caught yet, and the skies will fall in the time I turn to bay."

This is the same Voice that spoke to Job from the whirlwind. There are truths about existence as far beyond our feeble brains as our knowledge of the world is beyond the mind of a robin. Today Sunday could have shouted: "You may learn that matter is made of particles that in turn are made of superstrings, but that will not tell you why there are superstrings. If you ever succeed in reducing physics to a single equation, or a small set of equations, you will still not know the reason for those equations. You will never be able to explain why there is something rather than nothing, or why, as Stephen Hawking recently put it, the Universe 'bothers to exist.' "

"Listen to me," Syme cries in chapter 14. "Shall I tell you the secret of the whole world?" His speech wonderfully capsules the heart of G. K.'s nightmare as well as the heart of Plato. "It is that we have only known the back of the world. We see everything from behind, and it looks brutal. That is not a tree, but the back of a tree. That is not a cloud, but the back of a cloud. Cannot you see that everything is stooping and hiding a face?"

Nature has two sides, a front and back, and all of Nature is the back of God. In Plato's famous analogy, we see only shadows on the wall of the world's cave. In the boundless realm of all there is, beyond the fields we know, lies our only hope of escape from ultimate despair and death. At the close of Chesterton's nightmare, when Sunday begins to merge with God, he is able to call himself the Sabbath, the day on which God rested, the final peace of God.

In Exodus 33:20–23, God says to Moses:

> Thou canst not see my face: for there shall no man see me and live. And the Lord said, Behold, there is a place by me, and thou shalt stand upon a rock: And it shall come to pass, while my glory passeth by, that I will put thee in a cleft of the rock; and will cover thee with my hand while I pass by: And I will take away mine hand, and thou shalt see my back parts: but my face shall not be seen.

We know that Chesterton knew these verses. In his introduction to *The Man Who Was Thursday*, Wills calls attention to a passage in G. K.'s book *G. F. Watts* (1904) in which he speaks of Watts's unusual interest in painting human figures from behind. Here is what Chesterton writes on pages 62–63:

> Before we quit this second department of the temperament of Watts, as expressed in his line, mention must be made of what is beyond all question the most interesting and most supremely personal of all the elements in the painter's designs and draughtsmanship. That is, of course, his magnificent discovery of the artistic effect of the human back. The back is the most awful and mysterious thing

> in the universe: it is impossible to speak about it. It is the part of man that he knows nothing of; like an outlying province forgotten by an emperor. It is a common saying that anything may happen behind our backs: transcendentally considered the thing has an eerie truth about it. Eden may be behind our backs, or Fairyland. But this mystery of the human back has again its other side in the strange impression produced on those behind: to walk behind anyone along a lane is a thing that, properly speaking, touches the oldest nerve of awe. Watts has realized this as no one in art or letters has realized it in the whole history of the world: it has made him great. There is one possible exception to his monopoly of this magnificent craze. Two thousand years before, in the dark scriptures of a nomad people, it had been said that their prophet saw the immense Creator of all things, but only saw Him from behind. I do not know whether even Watts would dare to paint that. But it reads like one of his pictures, like the most terrific of all his pictures, which he has kept veiled.

"Sunday," Wills adds, "is Chesterton's attempt to paint that picture." Syme says it this way:

> "Then, and again and always," went on Syme, like a man talking to himself, "that has been for me the mystery of Sunday, and it is also the mystery of the world. When I see the horrible back. I am sure the noble face is but a mask. When I see the face but for an instant, I know the back is only a jest. Bad is so bad, that we cannot but think good an accident; good is so good, that we feel certain that evil could be explained."

If Chesterton had ended his fantasy with Syme's outburst about Nature as God's backside, his book would have been no more than an apology for philosophical theism, unlinked to any religious creed. But he does not stop there. The book closes with a dream sequence, a dream within a dream, involving a great costume ball in which the six policemen are clothed in ways that resemble the first six days of Genesis. Gregory, the book's authentic anarchist, becomes a symbol of Satan, the supreme destroyer. After he and Syme cross verbal swords, Syme denies

Gregory's charge that humanity has not suffered. Turning to Sunday, whose face wears a strange smile, he asks, "Have you ever suffered?"

Sunday's face expands until it fills the sky. Everything goes black. Nature, God's back, blurs and vanishes. God's transcendent face no longer can be seen. Only his voice can be heard as he asks, "Can ye drink of the cup that I drink of?" It is the only passage in the book from the New Testament. Chesterton's nightmare ends with a reference to the Incarnation—God taking a human form to experience human pain and to prepare for our eternal life. This is what Syme calls the "impossible good news," the gospel that makes "every other thing a triviality, but an adorable triviality."

Chesterton's point is this. If one manages that mysterious leap of faith, one escapes in the only possible way from what Miguel de Unamuno called the "tragic sense of life." Nature's evil backside fades in the light of God's peace. Commonplace things such as lampposts, apple trees, windmills, balloons, ships, hornbills, elephants, the moon—one of Chesterton's essay collections is titled *Tremendous Trifles*—take on a kind of gaiety they never had before. "About the whole cosmos," Chesterton wrote in *Heretics* (1905), "there is a tense and secret festivity—like preparations for Guy Fawkes day. Eternity is the eve of something." No longer are objects things that will forever be lost to our experience. Everything is seen with a new sense of wonder and gratitude. It would be many years before Chesterton could believe that the Good News was embodied and preserved by the Roman Catholic Church. He was received into the Church in 1922. His wife followed four years later.

Garry Wills in his introduction suggests that Chesterton intended *all* of Syme's wild adventures, not just the fancy dress ball, to be one of Syme's dreams. After all, on the title page of his novel he called it a nightmare. Syme's dream begins in the first chapter, when G. K. writes, "What followed was so improbable, that it might well have been a dream." Syme leaves the deserted garden party "with a sense of champagne in his head" to walk into a starlit street, where he encounters Gregory. The nightmare ends on the book's next-to-last page, when Syme, having no sense of awakening, finds himself strolling along a country lane with Gregory. They have talked all night, and morning is breaking. In chapter 2 Syme remarks, "I don't often have the luck to have a dream like this." Gregory replies, "You are not asleep, I assure you." But this may be only a dream Gregory speaking.

Mention should be made of Chesterton's usual swashbuckling style, with its dazzling metaphors and constant alliteration. Alliteration always came to Chesterton as easily as breathing. Most often it involves a pair of words with the same first letters, but at times there are three or even more such words close together. If one were to go through the novel and underline every case, I would guess there to be hundreds.

17

Ernest Hemingway and Jane

Jane
Said: "I'd be insane
To shoot my poor husband in the head
When I can shoot Ernest Hemingway instead."
—a clerihew by ARMAND T. RINGER

A shorter version of this article appeared in *The Skeptical Inquirer* (November/December 2001).

Chris Coover writing on "A Hemingway Discovery" in *Christie's* magazine (May/June 2000), reported on a recent find of Hemingway manuscripts, letters, and book galleys in a trunk stored by Jane Kendall Mason, an American socialite who had been one of the many women with whom Hemingway had romantic romps. After Hemingway lost interest in Jane, he used her as a model for the character of Margot in one of his most famous stories, "The Short Happy Life of Francis Macomber."

The trunk contained twenty-three letters from "Papa" to Jane, and a handwritten first draft of the short story in which Jane appears. In an act of cruelty, Hemingway had sent the draft to Jane before its final version appeared in the September 1936 issue of *Cosmopolitan*. She would, of course, have at once recognized herself as Margot, the adulterous wife who shoots her husband on a safari in Africa. The trunk also contained an unfinished short story by Jane, with Hemingway's revisions, titled "A High Windless Night in Jamaica."

My second excuse for writing about Jane is the little-known fact, which only recently came to light, that in her elderly years she became convinced she was possessed by demons, and actually underwent an exorcism by Eileen Garrett, a well-known New York City medium. Jane had become a Spiritualist and medium, producing thousands of pages of automatic writing by a hand she believed was being guided by discarnates. I will tell the wild story of her failed exorcism, but first an account of her flamboyant, frustrated life. For its details I rely mostly on "To Love and to Love Not," a remarkable article by Jane's granddaughter Alane Salierno Mason, a book editor at W. W. Norton, that ran in the July 1999 issue of *Vanity Fair*.[1]

Jane and Hemingway first met on an ocean liner in 1931. He was thirty-two and married to Pauline, his second wife. Jane was twenty-two and married to Grant Mason, a wealthy American who worked in Cuba as an executive for Pan American Airways. They lived near Havana in a large villa where Jane, who aspired to be a sculptor, had a third-floor studio.

An adopted daughter of Maryland's multimillionaire Lyman Kendall, Jane was often in the news as a prominent society woman of extraordinary beauty. An ad for Pond's Cream described her as "clean cut as a cameo in her Botticelli beauty of pale gold hair and wide-set eyes like purple pansies." Tall and athletic, Jane rode horses, fished, hunted in Africa, played the piano and harp, ran a shop in Havana to sell Cuban art, spoke three languages, and gave fabulous parties featuring pigeons dyed different colors and fresh flowers sewed to tablecloths. Among her many famous friends of later years were Dorothy Parker, Robert Benchley, Archibald MacLeish, and Clare Boothe Luce.

In 1932 Hemingway, leaving Pauline at their home in Key West, spent two months in Havana where he and Jane fished for marlin and became better acquainted. He once boasted that Jane had climbed through his hotel bedroom transom to get in bed with him.

[1] Alane Mason is the adopted daughter of one of Jane's two adopted sons. It was she who discovered the trunk mentioned in my opening paragraph. The draft of "The Short Happy Life . . ." was sold by Christie's to a private collector, his name not disclosed, for the highest price ever paid for the manuscript of a short story by an American author.

A car accident injured Jane's back. While recuperating from an operation in a Manhattan hospital, her husband sent her gifts from Havana which, much to the hospital staff's annoyance, included a tiny green monkey and two white mice named Samson and Delilah.

Jane further damaged her spine in a suicide effort by jumping from a low balcony. Hemingway, whose ego was immense, said to friends that she had tried to kill herself over unrequited love for him. He told John Dos Passos that Jane had literally "fallen" for him! In one of his letters he referred to her husband as a "twerp."

Jane had numerous love affairs, one with a big-game hunter named Dick Cooper. On an African safari with Cooper she managed to kill several lions and the foal of a rare white zebra belonging to an endangered species. The zebra's skin was sent to England to become a rocking horse for her two adopted sons. Jane later wrote a play, never produced, called *Safari*. It was about a woman named April who wanted to marry a South African army captain but hesitated because it might damage her social standing back home.

In Manhattan Jane was psychoanalyzed by Dr. Lawrence Kubie, a prominent Freudian. He said she was the only patient he ever had whom he couldn't help. Dr. Kubie wrote an article about Jane and Hemingway, which he sent to the *Saturday Review*. MacLeish persuaded the magazine not to print it because it libeled Hemingway by intimating, what everybody who knew him surmised, that "Hem" (as friends called him) suffered from deep doubts about his manhood—doubts that explained his mania for such macho interests as boxing, bull-fighting, hunting, fishing, boozing, warring, and womanizing.

The single word that best describes Hemingway's character is "bully." He relished picking fights with strangers, then suddenly, without warning, hitting the unguarded man on the jaw, often knocking him out. One of his sickest letters brags about how he beat up Wallace Stevens so severely that the poet was taken to a hospital.

The most publicized of many such episodes occurred in the office of Maxwell Perkins, Hem's editor at Scribner's. Furious because his old friend Max Eastman had written that Hemingway's writing suggested he had false hair on his chest, Hem smashed a book in Eastman's face. The thin-skinned Hemingway had taken Eastman's remarks to imply sexual impotence. Anticipating a left hook, Eastman, who had skill as a wrestler, pinned Hemingway's back on a desk and forced his head to the floor before office staffers rushed in and separated them. You'll find a hilarious account of this fracas in "The Great and the Small in Ernest Hemingway," a chapter in Eastman's *Great Companions*. The press had a field day with the story. In vain efforts to counteract his everlasting embarrassment, Hemingway gave several spurious accounts of the incident in which he claimed he restrained himself so as not to hurt Eastman. Hem never forgave Max. In one of his letters (I quote from John Updike's review of a collection of Hemingway correspondence) he wrote:

> I've got one ambition. Not an obsession. Just an ambition to nail that son of a bitch max eastman to the top of a fence post with a twenty penny spike through the base of his you know what and then push him backwards slowly. After that I'd start working on him.

Hemingway enjoyed killing humans as much as he enjoyed killing jungle beasts. Updike quotes from a disgusting letter Hemingway sent to Charles Scribner:

> One time I killed a very snotty SS kraut who, when I told him I would kill him unless he revealed what his escape route signs were said: You will not kill me, the kraut stated. Because you are afraid to and because you are a race of mongrel degenerates. Besides it is against the Geneva Convention.
>
> What a mistake you made, brother, I told him and shot him three times in the belly fast and then, when he went down on his knees, shot him on the topside so his brains came out of his mouth or I guess it was his nose.

This may be fantasy, Updike conjectures—Hemingway's third wife, Martha Gelhorn, said he was the biggest liar since Baron Münchhausen—but other letters brag about other killings of enemy soldiers in World War II. Racist remarks also pepper his epistles. There are endless references to kikes, wops, and jigs. Seldom has there been a man so bisected into a fine, sensitive writer, and a person whom Andy Rooney (in his July 3, 1999, column) called a "Big Jerk."[2]

Over and over again Hemingway would turn against former friends and sadistically bash them in books and letters. His volume of Paris memoirs, *A Moveable Feast*, contains cruel recollections of F. Scott Fitzgerald, Gertrude Stein, Ford Madox Ford, and others. This also happened to Jane. In a letter to MacLeish, Hemingway calls her a "bitch," adding that he would like to give her a burst of gunfire. He told one of his biographers that Jane was the "worst bitch" he had known at the time, and that her sole virtue was an eagerness to get laid.

After divorcing Grant, Jane married John Hamilton, a Republican bigwig. The marriage didn't last long. Following a torrid affair with Paul Palmer, a *Reader's Digest* editor, she married George Abell, European bureau chief for Time-Life, and a popular columnist. Divorced again, she married her fourth and final husband, Arnold Gingrich, founder and editor of *Esquire* and *Coronet*. Hemingway was flabbergasted when he learned of this marriage: "I can't get over it," he said. "I can't believe she married that little "t—."

In three of his works Hemingway based an unpleasant character on Jane. She is Helène Bradley in *To Have and to Have Not*. Gingrich, then a Scribner's editor, recognized Jane as the model for Helène, and believed many passages were libelous. Hemingway was furious when Gingrich insisted he remove the passages. Jane is also Dorothy, a stupid, spoiled, oversexed young woman in Hemingway's play *The Fifth Column*. And she is Margot in Hemingway's familiar story "The Short Happy Life of Francis Macomber."

[2] Andy Rooney tells how Hemingway, in a Paris hotel during World War II, got into a trivial argument with Bruce Grant, of the *Chicago Sun Times*. Suddenly the slightly built Grant found himself defending fisticuffs. I had the pleasure of meeting Grant at a gathering in Brooklyn where he gave a colorful account of the incident. He said he never expected Hemingway to floor him with a sudden jab to the jaw.

Andy considers a *New Yorker* story in which Hemingway describes how he and his wife killed an African lion. "Let's go," the wife says, sad because she hadn't killed the lion all by herself, "and when we're in bed we can listen to the night."

"That's not the night you hear," Andy ends his column, "that's the sound of a childhood Hemingway admirer throwing up."

In that story Margot is unhappily married to Francis, a wealthy but wimpy American whom she dominates. They hate each other. On an African safari, Macomber flees in terror from a wounded lion, making him a coward in his wife's eyes. Later, however, he suddenly loses all fear when he shoots at a buffalo charging toward him. Margot takes aim at the same beast but instead shoots her husband in the head. The story has a trick ending, like Frank Stockton's "Lady or the Tiger?" It is not clear whether Margot, sensing how her husband has changed, did or did not intend to kill him.

Here is what Hemingway said about the story in a letter:

> I wrote "The Short Happy Life of Francis Macomber" about a woman I was mixed up with one time who had a husband who was a coward. I knew he was a coward by direct observation and by local knowledge. But I invented the story in Africa instead of where it happened.

Though cleverly written, I consider this one of Hemingway's worst stories. Macomber's instant change from a coward to a brave man is much too improbable. I see the tale as just another one of Hemingway's not-so-subtle efforts to imply that he, The Great White Hunter, was a man without fear.

Hemingway died in 1961 at age sixty-two. He had become physically ill, severely depressed, and paranoid. Shock therapy was administered at the Mayo Clinic. I was walking east one afternoon along Forty-second Street in Manhattan, alongside the public library, when I passed Hemingway walking slowly the other way. He was staring straight ahead with a look of fear in his eyes. Soon after that, as everyone knows, his loudly trumpeted bravery deserted him. He put a double-barreled shotgun in his mouth and blew out his brains the way he once claimed he had done to a "kraut."

Jane's belief that she was possessed by spirits had its origin in a session with a Ouija Board when it began to glide under her fingers and spell out messages from the Great Beyond. When she began talking as if her mind and tongue were taken over by a demon, her husband Gingrich sought the help of Robert Laidlaw, M.D., who headed the

Psychiatric Department of Roosevelt Hospital. He in turn contacted his friend Eileen Garrett, who often assisted him in treating persons who fancied themselves possessed. Garrett was a famous Irish-born trance medium living in Manhattan. She had founded the Parapsychological Foundation in New York City, and *Tomorrow*, a magazine about the paranormal, which she edited with the help of Martin Ebon, her managing editor.

Jane's exorcism was witnessed and audiotaped by Ebon, then administrative secretary of the Parapsychological Foundation of which Garrett was president. A refugee at age twenty-one from Hitler's Germany, Ebon became a prolific writer of more than forty instructive and valuable books about paranormal topics, world communism and the Soviets, and numerous biographies of Soviet leaders. We have become good friends in spite of our opposing views about psi phenomena.

Ebon describes Jane's exorcism in "Ghost Against Ghost," the first chapter of his 1974 book, *The Devil's Bride: Exorcism, Past and Present.* However, names and details were altered, and it was not until 1999, when Ebon, speaking at the Parapsychological Foundation, disclosed that the person possessed was none other than Hemingway's former companion Jane.

Ebon calls her Victoria Camden. Her husband, Arnold Gingrich, is called Walter Camden. They are said to be living in a lavish town house on Manhattan's Upper East Side. Present during the exorcism were Jane, Garrett, Ebon, and Gingrich. Ebon was there to observe and audiotape.

For several years Jane had believed that her mind and body were repeatedly taken over by a variety of different spirits. One in particular claimed to be a Salem witch who had escaped detection and hanging. Ebon calls her Ruth, though actually she was nameless. She would fling Jane's body across a room and onto the floor. On one occasion, Jane said, the witch had almost drowned her in the bathtub.

Garrett went into her usual trance, and was first taken over by her major control, Uvani, a soldier who lived centuries ago in India. Uvani was then replaced by Abdul Lotif, a twelfth-century Arab physician, another of Garrett's controls. While in trance, Eileen's voice always changed markedly to the accents of the discarnate speaking through her.

As Ebon describes the scene, Jane began writhing with convulsions as she felt herself invaded by the witch. She fell to the floor, sobbing, then crawled over the rug to rest her head on Garrett's knees. Gingrich watched in stunned silence.

And now an incredible dialog took place. For the first time in the history of channeling, Ebon believes, a ghost argued with another ghost. Abdul did his best to persuade the witch to leave Jane alone. The witch refused.

At the end of the session Abdul lifted Garrett's hand until it rested on Jane's head. "And now, you," Abdul said, "must go and let this child reside in her own world. She must be restored to herself, and to herself alone."

Garrett groaned and shuddered as she came out of her trance. "What's happened?" she asked. Trance mediums almost never recall, or pretend not to recall, what they say while under a control. Jane slowly became herself. "I guess we all need a stiff drink," Eileen said.

While the group was having drinks and sandwiches, they discussed a male poltergeist that Jane thought had been making tapping sounds in the house.

Garrett assured Jane that the poltergeist was "a friendly spirit who likes the house, he likes you, but I've asked him to go away; to please go away in the name of God and leave everybody at peace until they are strong. I see him as brash, cheerful, nonchalant, good-natured but rough."

"Not too good-natured," said Jane.

Asked how she felt about her possession by the witch, Ebon quotes Jane as saying:

> I've suffered terribly with this, but I've never been afraid. Now that is the peculiar part. I don't understand it. You ride a horse that's thrown you and you may say to yourself, "I'm not afraid of this horse," but deep down in your soul, you are afraid but I was not afraid of this. I had some misgiving about coming back here tonight. I admit that. But still and all, when [Ruth] takes hold of me, as she did before, I'm still not really afraid of her, though I know she can hurt me.

The exorcism was only partially successful. Ebon tells me that for several months after the exorcism Jane was less persecuted by the witch. Jane's later trances seemed less genuine, more like theatrical performances to gain attention. Her case was complicated by severe alcoholism, which distorted and colored her thinking.

I tried to obtain Ebon's audiotape of the exorcism, but it seems to have been lost in the archives of the Parapsychological Foundation.

Jane always fancied herself a talented poet and novelist. Gingrich published some of her poems in *Esquire* under the pseudonyms of Proctor Farwell and James Matheson. After dying of cancer in 1980—her husband had died of cancer four years earlier—she left several unpublished manuscripts including a memoir of her childhood, a novel titled *Dear Meg*, and notes for a book about her experiences with demon possession. "Jane probably never really wanted to marry Ernest Hemingway," Mason writes. "She wanted to *be* Ernest Hemingway."

On Jane's tombstone, alongside Gingrich's, are words she herself wrote: "Talents too many, not enough of any." Mason closes the article about her grandmother by writing: "In the end she would not be remembered for her own talents, but for Hemingway's."

18

Three Parodies of Famous Poems

What's Done Is Done

(With apologies to Henry Longfellow)

The day is done and the darkness
Falls from the sky above,
As I sit in my lonely room dreaming
Of a long-ago hopeless lost love.

It's a feeling of sadness and longing
That is not akin to pain,
But resembles sorrow only
As the mist resembles rain.

Shall I call the ex-wife who divorced me
To come and hop into our bed?
No, she will only hang up on me.
I'll call on a call girl instead.

And the night will begin with our pleasures,
And the cares that infest the day,
At least for a couple of hours
Will silently steal away.

Jackson Pollock

(With apologies to Rudyard Kipling)

When Pollock's last picture was painted,
And his brushes were twisted and dried,
And his youngest critics were silent,
And his oldest detractor had died,
He shall rest, and faith he deserves it.
His dribblings are such a big bore.
It is said the best of his pictures
Were those that dripped off on the floor!

The Holdup Man

(With Apologies to Alfred Noyes)

The wind was a torrent of darkness
among the gusty trees.
The moon was a ghostly galleon
tossed upon cloudy seas.
The road was a ribbon of moonlight
where trucks and buses roar.
When the holdup man came driving—
driving—driving—
The holdup man came driving,
up to the old inn door.

Jack parked his car at the curbside,
then raced across the yard
To tap with his gun on the entrance

but all was locked and barred.
He whistled a tune at the window,
and who should be waiting there,
But the owner's black-eyed daughter.
Bess, the owner's daughter.
Adjusting some of the curlers
in her mop of jet-black hair.
"One kiss, my dearest sweetheart,"
but Bessie shouted, "No!"
"You lied when you swore you were single.
"You've a wife and four kids in Chicago.
"So hit the road, Jack, and don't you come back.
"I've finally gotten your number.
"Now I'm engaged and soon will be wed
"to Luke Warum, our city's top plumber."

"I'll kill you for that," snarled the felon.
"You must think I'm just a big sap."
But Jack didn't know that sly Bessie
was concealing a gun on her lap.
Two pistols went "bang!" in the darkness,
but Jack's was such a bad shot
That his bullet whizzed past Bessie's curlers.
and her ex-love expired on the spot.

19

Edgar Wallace and *The Green Archer*

Two curious mysteries relate to Edgar Wallace. One is why, with only a few mediocre exceptions, none of his books are in print today in the United States. His novels and short stories were enormously popular throughout the 1920s, both in England and here. In 1982, fifty years after his death, all Wallace's books became public domain in England, and most are now out of copyright in the United States. Yet when I checked the latest *Books in Print*, only five of his most forgettable novels were listed, one of them dreary science fiction. The other mystery, with which I conclude this chapter, concerns what I consider his single masterpiece, *The Green Archer*.

Wallace's adventurous youth and prodigious literary output are now legendary. His more than 170 hardcover books, some 950 short stories, 23 plays, and countless newspaper and magazine articles top even the productivity of Isaac Asimov. There was a time in England when one out of every four books sold, excluding textbooks and the Bible, was by Wallace. During one year he had three plays running simultaneously in London. Most of his plays had short runs, but some were great successes, notably *On the Spot*, said to have been written in four days. It starred Charles Laughton as a gangster based on Al Capone, and became a novel in 1931. Scores of Wallace's books were nonfiction, including three ghosted autobiographies and ten volumes on the history of the First World War!

So rapid was the speed at which Wallace dictated his fiction that rumors circulated about a crew of ghostwriters hired to write his novels after being given plot outlines. Wallace denied this vigorously. He offered five thousand pounds to anyone who could prove he did not write every sentence in his books. There were no takers. No ghostwriters have since surfaced.

Richard Horatio Edgar Wallace was born on April 1, 1875, in Greenwich, England. Both parents were actors and unmarried. At age nine he was adopted by George Freeman, a fish porter who named him Dick Freeman. Only in later years did Edgar learn that the Freemans were not his real parents.

As a boy Edgar had a variety of odd jobs. He sold newspapers at Ludgate Circus, where a bronze plaque now commemorates this fact. He worked for a printer, in a shoe shop, in a rubber factory, and as a milkman. As Howard Haycraft remarks in *Murder for Pleasure* (1941), a job as errand boy laid the foundation for his vast knowledge of London. He had no formal education beyond the age of twelve.

At eighteen Edgar enlisted as a private in the Royal West Kent Regiment. He served four years before being transferred to the Medical Staff Corps in South Africa. It was in Africa that he began writing verse in the manner of Rudyard Kipling, whom he greatly admired. When Kipling visited South Africa, Wallace welcomed him with a poem titled "Good Morning, Mr. Kiplin'." Wallace also wrote lyrics for a popular singer named Arthur Roberts. When the Army refused him leave to hear Roberts sing, he went anyway. Punishment was thirty-six hours of hard labor in an Army prison. Four collections of his verse were published: *The Mission That Failed: A Tale of the Raid and Other Poems* (Cape Town, 1898); *Nicholson's Neck* (Cape Town, 1900); *War! and Other Poems* (Cape Town, 1900); and *Writ in Barracks* (London, 1900).

After Wallace bought himself out of the Army in 1899, Reuters hired him as a war correspondent to cover England's notorious Boer War. It was in Africa in 1901 that he married Maud Caldecott, daughter of a South African missionary. In 1902 he edited Johannesburg's *Rand Daily News*. Back on London's Fleet Street he became a crime reporter for *The Daily Mail*. It has been said he was the first to notice that when a jury returned to the courtroom with a verdict, they looked at a prisoner only if they had found him innocent.

In 1904 Wallace became editor of *The Evening News*. A year later he founded the Tallis Press to publish a collection of short stories titled *Smithy* (1905) that had earlier appeared in *The Daily Mail*, and his first novel, *The Four Just Men* (1905). It was the beginning of a series about four men, later three, who took justice into their own hands by executing persons who had committed horrendous crimes but were beyond the reach of England's laws. To boost sales, Wallace offered five hundred pounds in prizes to readers who figured out how the book's murder was committed. The novel was a financial disaster.

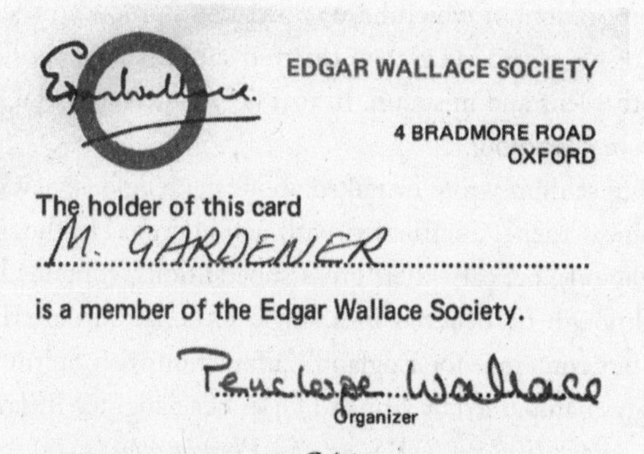

EDGAR WALLACE SOCIETY

4 BRADMORE ROAD
OXFORD

The holder of this card
M. GARDENER
is a member of the Edgar Wallace Society.

Penelope Wallace
Organizer

Renewal Date 1st January 19 84

My membership card, name misspelled

The Daily Mail sent Wallace to the Belgian Congo in 1907 to cover atrocities. His experiences there became the basis for his finest short stories. First published in periodicals, they later appeared in a series of eleven books starting with *Sanders of the River* in 1911. Perhaps, G. K. Chesterton observes in *Come to Think of It*, Wallace was a better writer before he became a best seller. *The Daily Mail* fired him after some of his dispatches embroiled the paper in a costly libel suit. For the next few years Wallace worked for many London papers, on several of which he served as editor.

Wallace and Ivy were divorced in 1918 after having three children: Bryan, Michael, and Patricia. Following a decade of loyalty to a mistress, in 1921 Wallace married his secretary, Ethel Violet King, twenty years his junior. They had one child, Penelope. After her father's death, Penny founded in 1969 the Edgar Wallace Society, of which I became a member. It published a quarterly *Edgar Wallace Newsletter*, a valuable source of information about Wallace and his writings.

Wallace's career as a mystery writer took off like a rocket in the late 1900s. He lived lavishly on royalties, much of which was squandered on racing bets. Wallace had a lifelong compulsion to gamble at the racetrack, and at one time owned twenty racing horses. He smoked constantly, keeping his cigarettes a foot from his nose in a long cigarette holder. He became chairman of London's Press Club. The British Lion Film Corporation, of which he was president, produced silent versions of eight of his plays. More than thirty-five of his later thrillers became films, both silent and in sound. In 1931 he was defeated as a liberal candidate from Blackpool.

Wallace seldom wrote or talked about his religious views. He was a philosophical theist, unaffiliated with any church. He thought that all major religions, beneath their gross superstitions, contained kernels of truth. Although he believed in a single God, and an afterlife, he had nothing but contempt for England's infatuation with Spiritualism, then so strongly championed by Conan Doyle. His daughter-in-law Margaret Lane, in *Edgar Wallace: The Story of a Phenomenon* (1948), quotes from one of Wallace's attacks on Spiritualism:

> I do not believe that anyone, . . . by falling into an epileptic fit or a good imitation of one, secures the mysterious power of bringing themselves into touch with these personalities which have no longer habitation in the human frame. . . . Why should spirits blow horns and tin trumpets, and pick up tambourines and shake them? When we depart this mortal life do the sanest of us become clowns? Why is all this dreary nonsense necessary if it is not that it is tricks which are

> easily performed by an unscrupulous medium? . . . As I have said before, charlatanism reigns in this peculiar sphere.

In his later years Wallace became as corpulent as Chesterton. He hated all exercise, including walking, and is said to have taken taxis in London rather than walk a hundred yards. A contract with RKO took him to Hollywood, where he worked on scenarios and wrote the script for *King Kong*. It was in Hollywood that a diabetic coma developed into pneumonia. He died in his sleep in 1932. When his body was returned to England, flags at the seaport were lowered to half mast.

It is not generally known that Wallace, like Conan Doyle, tried his hand at science fiction. You'll find summaries of his wild plots in Everett Bleiler's monumental *Science Fiction: The Early Years* (1995, pages 783–84). The novels are: "1925": *The Story of a Fatal Peace* (1915); *The Green Rust* (1920); *The Day of Uniting* (1926); *Planetoid 127* (1929); and *The Sweizer Pump* (1929). *The Black Grippe* (1920) was a short story. None of these works are worth reviving, although *The Green Rust* is currently available here as a paperback. *The Day of Uniting* is about a failed prediction that a comet is about to destroy the earth.

Numerous detectives, mostly from Scotland Yard, appear in Wallace's thrillers, some in more than one book, but far and away the most memorable of his sleuths is J. G. Reeder. His first appearance was in the novel *Room 13* (1924), next in a collection of short tales, *The Mind of Mr. J. G. Reeder* (1925), retitled in America as *The Murder Book of J. G. Reeder*. (Many of Wallace's books, to the confusion of bibliographers, had different U.S. titles.) Reeder reappears in the novel *Terror Keep* (1927), and *Red Aces* (1930), and in a story collection, *The Guv'nor* (1932), reprinted in the United States as *Mr. Reeder Returns* (1932).

Reeder resembles Chesterton's Father Brown, and Peter Falk's Columbo, in giving the impression of a meek, mild-mannered, incompetent person incapable of solving any crime. Here is how Wallace describes him in *The Mind of J. G. Reeder*:

> There was a certain wistful helplessness about John G. Reeder that made people feel sorry for him, that caused even the Public Prosecutor a few uneasy moments of doubt as to whether he had

been quite wise in substituting this weak-appearing man of middle age for Inspector Holford—bluff, capable and heavily mysterious.

Mr. Reeder was something over fifty, a long-faced gentleman with sandy-grey hair and a slither of side whiskers that mercifully distracted attention from his large outstanding ears. He wore halfway down his nose a pair of steel-rimmed pince-nez, through which nobody had ever seen him look—they were invariably removed when he was reading. A high and flat-crowned bowler hat matched and yet did not match a frock-coat tightly buttoned across his sparse chest. His boots were square-toed, his cravat—of the broad, chest-protector pattern—was ready-made and buckled into place behind a Gladstonian collar. The neatest appendage to Mr. Reeder was an umbrella rolled so tightly that it might be mistaken for a frivolous walking cane. Rain or shine, he carried this article hooked to his arm, and within living memory it had never been unfurled.

Why the umbrella? Because concealed in its handle was a knife.

Second only to the Sanders tales, Wallace's stories about Mr. Reeder are the most worth reading today of all Wallace's hundreds of short stories, many of which never found their way into books. As for his mystery novels, there have been numerous listings of first editions and reprints. The most accessible list is among the twelve pages devoted to Wallace in *The Dictionary of Literary Biography*, Volume 70.

With one exception, which I will come to in a moment, all of Wallace's thrillers were hastily written and can only be considered second rate. One must marvel, however, at the novelty of their plots, their unusual characters, and the swift pace at which their narratives move. *The Clue of the New Pin* is a locked-room mystery, an ordinary pin providing the only clue. *The Avenger* (retitled *The Hairy Arm* in the U.S.) is about a highly intelligent ape. *The India Rubber Men* tells of a gang of thieves who wear rubber shoes, rubber gloves, and rubber gas masks. *The Black*'s main character is a man who dresses in black and wears a black mask. Surefoot Smith is the detective in *The Clue of the Silver Key*. Betcher Long, so nicknamed because he likes to make wild wagers, is the sleuth in *The Terrible People*. *The Girl From Scotland Yard* features a female detective. Jane, in *Four Square Jane*, is a jewel thief who steals from the rich and gives to the poor. I was tempted to obtain a copy of *The Dark Eyes of London*, because of its poetic title, until I learned it was about a gang of blind men.

I turn now to what I earlier called the second mystery about Wallace. How is it possible that among his raft of thrillers one and only one towers far above all the others? In my opinion it is the only Wallace novel likely to survive the test of time. I refer to his 1923 gothic mystery, *The Green Archer*.

Wallace does not mention this novel in *People*, his autobiography. For reasons I cannot fathom, he seemed not to consider this book superior in any way to his other thrillers. Nor is it singled out as exceptional in any of several biographies of Wallace or in critical writings about his work. Yet *The Green Archer* is unlike all his other novels in the intricacy of its plot, and in the quality of the writing, above all in featuring men and women with richer personalities than the two-dimensional characters found in his other novels. Indeed, it is a work that rises almost to the level of a novel by Dickens.

The Green Archer's heavy is Abel Bellamy, a large, muscular, ugly-faced, evil man. With the wealth he has acquired as a Chicago building contractor, he has retired to a medieval castle that he bought in Berkshire County, England. Garre Castle, as it is called, is about twenty miles west of London. An old legend tells how the castle has long been haunted by the ghost of a man called the Green Archer. He had been hanged in 1447 for stealing deer from the castle's owner.

The novel opens when Spike Holland, a young red-haired American from New York, now a crime reporter for London's *Daily Globe*, is asked to investigate a rumor. A man dressed from head to foot in green, his face covered by a white mask, has been seen prowling about the castle's grounds.

The first murder occurs when Charles Creager, an unsavory associate of Bellamy, is found dead with a green arrow through his chest. I will not spoil the pleasure of anyone who has not yet read *The Green Archer* by going deeper into the book's complicated action. It's central mystery, unsolved until the book's end, is the identity of the Green Archer. Major characters are Jim Lamotte Featherstone, a Scotland Yard captain hired by Walter Howett to protect his beautiful adopted daughter, Valerie.

Howett is an Englishman who moved to America where he became wealthy when oil was found on his farm. Valerie is in Garre on a mysterious mission to find out what became of her real mother, Elaine Held. She has reason to suspect that her mother's disappearance many years ago is somehow linked to Bellamy.

Other characters include Julius Savini, an olive-skinned Eurasian from Portugal. Formerly a member of a gang of thieves, he is married to Fay Clayton, a pretty, clever young woman with yellow hair and brown eyes who also has a criminal past. Coldharbour Smith, another evil associate of Bellamy, owns a club in Limehouse called The Golden East. It is a disreputable bar and dance hall, near the Thames. John Wood is a tall, handsome, modest bachelor who runs a home in Belgium for consumptive children.

The novel's improbable plot bristles with suspense and surprising revelations. Its chapters, first serialized in a periodical, often end in cliff-hangers. Wallace was not a great novelist, but he was a great storyteller and he makes it easy for readers to suspend disbelief. Once started, the book is difficult to put down. A final scene has Featherstone and Valerie, along with Savini and his wife, and a criminal called Lacy, trapped in one of Garre Castle's underground dungeons. Bellamy is trying to drown them by flooding the dungeon with water. They are released at the last minute, the water lapping their chins, by the Green Archer. Wallace liked to steal such dramatic scenes from himself. In the last chapter of *The Mind of J. G. Reeder*, Mr. Reeder and a young woman are similarly locked in a room with rising water.

Wallace's narrative is so carefully constructed that one must read the book a second time to fully grasp all its subtleties. After learning the Archer's identity, earlier scenes and conversations reveal fresh insights into the Archer's character and abilities. For example, in chapters 36 and 59 the Archer answers a series of questions so adroitly that he manages to conceal the truth without actually lying. His high intelligence is combined with a sense of humor that helps explain his whimsical masquerade as a ghost. Why arrows instead of bullets? Because a shot arrow makes no noise, and homemade arrows are untraceable.

In chapter 17 we learn that a certain wall is so high that one must extend hands upward as far as possible to touch its top. Many chapters later, the man we do not yet know as the Archer easily scales the wall without a ladder. Wallace never describes the Archer as muscular, but this is one of many indirect hints of his strength and athletic agility. On a first reading it is also easy not to realize that the Archer is the man casually mentioned in chapter 51 as slowly rowing a boat up the Thames in a thick, yellow fog.

In the second of the Archer's murders, which really are justified executions, a steel-tipped green arrow not only penetrates Coldharbour's chest, but is embedded in the wooden back of the chair in which he sat. One thinks of that final scene in Homer's *Odyssey* when a disguised Ulysses, after Penelope's wicked suitors have failed to bend her husband's old bow, bends it easily. When Bellamy is finally slain by the Archer, *two* green arrows penetrate his heart before he thuds to the floor.

Chapter 69, "The Man From Cloister Wood," is strangely effective. It follows the Archer's actions without revealing his identity, even though he is not wearing his green attire and white mask. One has the curious feeling of watching a scene, as if on a movie screen, with the man's face somehow concealed.

Wallace never wastes time with long descriptions of weather or scenery beyond a crisp sentence that often is beautifully worded. I give two examples. "Rain was now falling heavily and a chill wind swept bleakly through the square" (chapter 13). "The world was very quiet and beautiful, the half-moon turned a distant river to silver, and painted the park in soft, elusive tones" (chapter 26).

The novel has its blemishes. Bellamy's bestial qualities are overdone, although as in all novels of revenge it makes his execution all the more satisfying. Valerie has a tendency to blush and faint too easily. Racist words offend today's ears, and British slang expressions are meaningless to American readers. There are a few careless inconsistences. The Archer's hands were called white in chapter 3. In chapter 71 they are "brown, sinewy." As in all Wallace fiction there are clichés common to later popular fiction, but one must remember that they were not clichés when Wallace invented them.

The phrase "caught his man scientifically" (chapter 49) makes no sense. Perhaps "simultaneously" was intended. I suspect a printer's error. In the American edition "air" (page 190) is a misprint for "hair."

Creager's first name is called Charles in chapter 5, but he signs a letter "J. Creager" in chapter 15. In chapter 62 an unmade bed is called "discovered." This may have been a British synonym for "uncovered."

The Green Archer has twice been a motion picture serial. The silent version, in ten episodes, was produced by Pathe in 1925, starring Walter Miller as Featherstone and Allene Ray as Valerie. Burr McIntosh was Bellamy. I can still recall my excitement as a boy of eleven in Tulsa when I eagerly awaited next week's episode. After the serial ended I asked the theater's manager to let me have a dozen of the still photos, printed in green, that were displayed outside the theater. Alas, I lost them long ago. I was so taken by the serial that I had my parents buy a copy of the American hardcover, published in Boston, 1924, by Small, Maynard and Company. I have since read it several times, always discovering something I had not noticed before. It recently occurred to me that the little gold statue of Pan, on the Archer's desk, suggests that he is some sort of pagan in his religious beliefs. Vincent Starrett, by the way, wrote the introduction to a 1965 Norton reprint, although he writes only about Wallace, and says nothing about the novel.

In 1940 *The Green Archer* was again serialized, in sound of course, with fifteen black-and-white episodes. Never has a book's plot been so mangled. There is no Featherstone. Victor Jory took the role of Spike Holland, a detective not a reporter. His love, Valerie, played by Iris Meredith, is in London seeking a lost sister. In 1961 a third filming of *The Green Archer* was made in Germany. A two-act stage dramatization by Tim Kelly was published in 1980 by Barker's Plays, Boston. I have no information about its many amateur productions.

The Green Archer was first serialized in England's *Detective Magazine*, July 1923 through January 1924. I tried to get copies from the British Library, but their run had been destroyed during World War II. Oxford's Bodleian Library has a bound volume, but its pages are too fragile to permit photocopying. It is always interesting to compare a magazine text with the text of a book. Its first edition was published in 1923 by Hodder and Staughton. The same British firm issued a paperback in 1981 in which many paragraphs were cut. The Norton edition also had omissions. Did the magazine pages have illustrations, I wonder?

It's a shame that no edition of Wallace's one great classic thriller is now in print. Maybe someday a publisher will allow me to edit an *Annotated Green Archer*.

This is the poem that Private Wallace wrote to honor Kipling's visit to South Africa. His meeting with Kipling, who praised his verse, was a highlight of Wallace's army career.

Good Morning, Mister Kiplin'

O, good mornin', Mister Kiplin'! You are welcome to our shores:
To the land of millionaires and potted meat:
To the country of the "fonteins" (we 'ave got no "bads" or "pores"),
To the place where di'monds lay about the street
  At your feet;
To the 'unting ground of raiders indiscreet....

We should like to come an' meet you, but we can't without a pass;
Even then we'd 'ardly like to make a fuss;
For out 'ere, they've got a notion that a Tommy isn't class;
'E's a sort of brainless animal, or wuss!
  Vicious cuss!
No, they don't expect intelligence from us.

You 'ave met us in the tropics, you 'ave met us in the snows;
But mostly in the Punjab an' the 'Ills.
You 'ave seen us in Mauritius, where the naughty cyclone blows,
You 'ave met us underneath a sun that kills,
  An' we grills!
An' I ask you, do we fill the bloomin' bills?

Since the time when Tommy's uniform was muskatoon an' wig,
There 'as always been a bloke wot 'ad a way
Of writin' of the Glory an' forgettin' the fatig',
'Oo saw 'im in 'is tunic day by day,
  Smart an' gay,
An' forgot about the smallness of his pay!

But you're *our* partic'lar author, you're our patron an' our friend,
You're the poet of the cuss-word an' the swear,
You're the poet of the people, where the red-mapped lands extend,
You're the poet of the jungle an' the lair,
  An' compare
To the ever-speaking voice of everywhere. . . ."

20

Afterword to *The Green Archer*

> Then, unexpectedly, he [Featherstone] saw an old friend [Fay Clayton], and instantly all thoughts of Valerie vanished from his mind.
>
> —*The Green Archer,* Chapter VIII

The Green Archer is a tidy little world, its absurd plot made credible by the great narrative skill of Edgar Wallace. To anyone who has read this complicated novel carefully, it is apparent that next to John Wood, Fay Clayton is the story's most impressive character. Valerie Howett, though college trained, rich, and beautiful, seldom makes a memorable remark. She blushes too easily, faints or almost faints at the slightest provocation. You can't imagine Fay fainting under any circumstances. It was stupid of Valerie to suspect that her foster father, with his weak eyes, or Jim Featherstone, a high official at Scotland Yard, could possibly have been the Green Archer.

Fay, uneducated but streetwise, a former confidence woman who had three times been in prison, is much brighter than Valerie and almost as pretty. She is clever, good-natured, easygoing; her remarks are often witty, occasionally peppered with colorful metaphors such as the time she likened her frightened husband to jelly shaking in an earthquake. Here are my best guesses at what happened to the book's major characters after the end of the novel. If you have not had the pleasure of reading *The Green Archer*, you may skip what follows as meaningless, and end the chapter after this sentence.

Mr. Howett, Elaine Held, and Featherstone all settled in Cleveland. Elaine turned down Howett's marriage proposal. She rented an apartment near the large house Howett bought for Valerie and Jim. It was a happy marriage for the first few years, then ugly differences began to emerge.

Jim, who easily found work as a detective with the local police, wanted children. Valerie did not, and Jim reluctantly agreed. Valerie became Cleveland's most glamorous socialite. Her days were spent playing golf, taking piano lessons, sailing her yacht, learning French and Italian, riding her horses, and attending and giving lavish parties. Jim, like Sherlock Holmes, loathed society, especially Cleveland society. Valerie came to believe she had married far beneath her. It was hard not to wince when she had to tell someone her husband was a city cop.

More and more the pair drifted apart. Eventually they were divorced. Long homesick for London, Jim returned to the big city, and was soon back at Scotland Yard. Late one Sunday afternoon, on a sunny spring day, Jim was cutting across the northeast corner of Hyde Park when he recognized an old friend. Fay Clayton was sitting on a bench reading a paperback.

"Fay! Fay!" Jim shouted. "Is it really you?"

"Hello, Feathers. Yes, it's me. Or should I say it is I?"

Fay put the book aside, marking where she stopped reading with a business card. She looked radiant, with her makeup skillfully applied, wearing a fashionable dress, and her yellow hair down to her shoulders.

"Sit down," Fay said, moving to one side of the bench. "I hear you're back at the Yard. The scuttlebutt there is that you and Valerie have split. True?"

"I'm afraid so, Fay. But who told you that?"

"Oh, I have my spies. But I'm not surprised. Valerie had you bamboozled. She was a bit too—how shall I say it?—a bit too refined and rich for a Scotland Yard bum."

"Maybe you're right. Someday I'll tell you all the sordid details. Now it's your turn. How go things with you and Julius?"

Fay's face clouded. "I thought I loved the little guy. I really did. But I was very young and naive, and mixed in with bad companions. Julius was smart in many ways, but in other ways he was a stupid fool. He got involved with a gang of idiots who thought they had figured out a new and clever way to steal the Crown jewels. I did all I could to persuade Julius that the scheme wouldn't work, but he wouldn't listen. He wanted to make one final big haul that would put us on easy street for life. Poor Julius! He was shot and killed by one of the guards."

"How terrible! I'm truly sorry, Fay. Did the police think you were involved?"

"For a while they did. But I had left Julius before the plan was attempted. I never knew its details or when it was supposed to take place. The Yard gave me a hard time, but finally let me go."

"You were lucky, Fay. You might have done time for not reporting the plan."

"I know. I told the Yard I didn't think Julius was serious. They finally believed me."

"I always liked Julius, especially after he saved my life in those watery dungeons of Garre Castle."

Fay shuddered with the memories.

"Have you remarried?" Jim asked.

Fay shook her head. "I sold our egg farm for a hefty profit, and used the money to take courses in literature, art, and jewelry design. And how to speak like a lady. Have you noticed? I own a little shop off Picadilly. It sells jewelry, most of it my work. You must visit it sometime."

Jim was silent for a moment, then finally said, "Fay, I recall well it was in this very park, near the Marble Arch, that we ran into each other ages ago. You lied when you denied you were married. Remember? Then your big brown eyes looked straight into mine and you said—I still remember your exact words—'No, I'm not married, though I don't know what would happen if you pressed me very hard. I've always had a weakness for the pretty-boy type. What do you say, Featherstone?' "

Fay laughed. "You're not so pretty now, Feathers. But I must admit that age has improved your looks."

"I can say the same about you, Fay. You really look marvelous. But when you told me you had a weakness for the pretty-boy type, I knew you were joking."

"*Au contraire*. I was half serious."

Jim's eyes brightened. "In that case would a former felon object if this no-longer pretty boy did a little pressing?"

"You can press as much as you like," Fay replied with a faint smile. "In fact, if you don't object, *I'll* start the pressing."

Fay turned on the bench, put her hands on Jim's broad shoulders, then pressed a firm kiss on his lips. It caught Jim completely by surprise. He could smell Fay's subtle perfume and the even better fragrance of her gold hair as it swirled across his cheeks.

Struggling to think of what to say, Jim finally stammered, "I'd . . . I'd offer you a cigarette now, but I stopped smoking years ago."

"Me too," Fay said with a smile. "I used to chain-smoke, so it wasn't easy. But I finally made it without getting plump or hooked on booze."

Still a bit flustered, Jim glanced at his watch, then looked westward where a glittering sun, low on the horizon, was coloring the sky. "It's getting late. May I take you to dinner?"

"No, no! We'll dine at my place."

A few months later Fay and Jim were engaged. Before the year ended they were married in a civil ceremony by a judge Jim knew, and living in a modest flat midway between Fay's shop and Scotland Yard. Soon Fay was expecting her first child.

The Green Archer's identity was never disclosed by those who knew. Not even Spike Holland knew. Tired of London, he returned to America where he became a top crime reporter for the *Chicago Tribune*.

John Wood moved his school for consumptive children to Switzerland as he had planned. After Mr. Howett died, Valerie and her mother made frequent trips to London, Paris, Rome, and the Riviera. Wood kept in touch with his sister and mother by correspondence, and twice joined them in Paris. They marveled at the ease with which he spoke French to the natives.

Valerie finally married a wealthy Cleveland attorney. She made no efforts to contact Jim and Fay on any of her trips to London.

Garre Castle was taken over by the Crown to become the Green Archer Art Museum. Its paintings include, of course, all those owned by

Abel Bellamy. On one wall are the words "The Mystery of the Green Archer." Below are framed news accounts written by Holland, alongside photos of the principle characters involved. One of the Archer's green arrows is mounted on the wall under glass.

There is no mention or photograph of John Wood.

21

The Tin Woodman of Oz

Queer things happen in the Land of Oz.

—POLYCHROME in *The Tin Woodman of Oz* (1918)

The Baum Bugle, Fall 1996

In Baum's first Oz book, the Tin Woodman (his former name is not given) was at one time a man of flesh and blood. As he tells Dorothy, he had fallen in love with a Munchkin girl (also unnamed) who worked for an Old Woman who did not want to lose her as a servant. To prevent the woodman from marrying the girl, the Old Woman paid the Wicked Witch of the East—the witch destroyed by Dorothy's house—to enchant the woodman's axe, causing it to periodically hack off parts of his body. A clever tinsmith replaced each lost part with one of tin until eventually the woodman was made entirely of tin. (We are not told what happened to the original parts.)

Lacking a heart, the Tin Woodman no longer loved the Munchkin maiden. Caught in a rainstorm, his joints rusted, and he remained immobile for a year until Dorothy and the Scarecrow rescued him.

In *The Tin Woodman of Oz*, we learn that although the tin man now has a kind heart, it is not a loving heart. Perhaps the Tin Woodman did not earlier realize this because in *Dorothy and the Wizard in Oz* he thanks the Wizard for giving him a heart that "beats as kindly and lovingly today as it ever did."

It is Woot the Wanderer who convinces the tin man that even though he cannot love Nimmie Amee (her name now revealed), his heart is kind, and the kind thing to do is to "follow the bugle call of duty" and make good on his promise to marry her. The Scarecrow agrees, adding that this could be "best for the girl, for a loving husband is not always kind, while a kind husband is sure to make any girl content." Together with Woot, the Tin Woodman and his best friend, the Scarecrow, set out on their quest for Nimmie. The tin man hopes to marry her and make her empress of the Winkies.

After a series of adventures along the way, the book comes to so gruesome a climax that I suspect most children are deeply disturbed by it. It frightened me when I was a small boy. That may explain why I have not looked into the book until I reread it—with high pleasure—for writing this essay.

The Tin Woodman's account of how he became tin, as he tells the long story to Woot, differs slightly from the account he earlier gave to Dorothy. There is no Old Woman. Nimmie Amee is now a servant of the Wicked Witch. At the close of the book, Nick Chopper,[1] for that was his former name, discovers that the tinsmith, whose name is Ku-Klip, has preserved his former head in a cupboard. During a bizarre dialogue with his head, the Tin Woodman is understandably mystified: Is he the tin man, or is he the head in the cupboard? "Good gracious!" he exclaims. "If you are Nick Chopper's head, then you are *me*—or I'm you—or—or—or—What relation are we, anyhow?"

Before this conversation occurs, we learn of a series of unbelievable coincidences. After Nick deserted Nimmie, she fell in love with a soldier

[1] Fred Meyer has informed me that the name Niccolo Chopper first appeared in Baum's musical version of *The Wizard of Oz*. In *Lucky Bucky in Oz*, Fred also tells me, Nick speaks of six nieces, so he must have had a brother or sister. Each niece we learn, married a tinsmith. In chapter 5 of *The Wonderful Wizard of Oz*, the Tin Woodman says that after his father died, he took care of his "old mother as long as she lived." The Royal Historian wrote this, of course, before he learned that no one ages and dies in Oz.

named Captain Fyter. The Wicked Witch again works her magic by enchanting the soldier's sword so that he suffers the same horrible fate as Nimmie's former lover. The sword gradually slices his body to pieces. Once more it is Ku-Klip who replaces each missing part with tin. Apparently the soldier's tin heart is capable of love because he still loves Nimmie. Alas, on his way to elope with her, he is caught in a rainstorm. Like the Tin Woodman, he is rendered immobile by rust until he is rescued by the Tin Woodman, the Scarecrow and Woot. The two tin men now join forces to locate Nimmie Amee and ask her to choose between them.

Nimmie is eventually found in her magically fortified house at the foot of Mount Munch, a steep mountain inhabited at the top (as we learn in *The Magic of Oz*) by the Hyups. Nimmie is happily married to a surly creature called Chopfyte. Ku-Klip, having acquired a magic glue, has glued together the abandoned parts of both men to make a composite man of meat. The Tin Soldier recognizes his former head on Chopfyte. The Tin Woodman recognizes his former right arm by two warts on the hand.

Who is Captain Fyter? Is he Chopfyte because the scowling misfit has his meat brain? Or is he the reconstituted man of tin with tin brains and a tin heart?[2]

Nimmie Amee is surely one of the strangest meat people in Oz. Although so beautiful that she makes sunsets blush, and presumably capable of finding excellent meat suitors, she much prefers men of tin! Would not such a man be "the brightest of husbands?" She won't have to cook for him because he never eats, or make his bed because he never sleeps, or find him tiring at dances. "I shall be able to amuse myself in my own way—a privilege few wives can enjoy." Moreover, Nimmie says,

[2] Plutarch describes what philosophers call the paradox of the *Ship of Theseus*. Over decades, parts of a ship are gradually replaced by new parts. This occurs in such small increments that sailors never doubt they are living on the same ship. Imagine that the old parts are preserved and later reassembled. Which is now the "real" ship? For a discussion of similar paradoxes of identity, especially involving humans, see chapter 19 of my book *Whys of a Philosophical Scrivener* (William Morrow, 1983).

Baum had a strong interest in theosophy and was a firm believer in reincarnation. This may have heightened his fascination with what I like to call the "Tin Woodman problem." If in our next incarnation we have a completely different body and brain, in what sense are we the same person who lived earlier?

"I shall take pride in being the wife of the only live Tin Woodman in all the world." These sentiments, Baum adds with sly sarcasm, show that Nimmie Amee was "as wise as she was brave and beautiful."

When Captain Fyter is sliced to pieces and rebuilt with tin, Nimmie at once agrees to marry him. Again, she is disappointed when the soldier fails to show up for their wedding. When Chopfyte arrives on the scene, she likes him because one of his arms is made of tin, and he reminds her of her two former suitors!

Although *The Tin Woodman of Oz* may be the least satisfactory of Baum's Oz books, it is filled with familiar delights of whimsy, humor, and wordplay. In Loonville, the inhabitants are puffed-up, purple-eyed balloons. Bal Loon is obviously "balloon." Panta Loon is "Pantaloon," a buffoon character in British harlequinades (pantomime comedies) as well as a type of trousers. In Baum's original manuscript, Til Loon is called Sal Loon, a play on "saloon." Evidently the publisher did not think this word suitable for a children's book at a time when the temperance movement was cresting.

Did Baum have any word play in mind when he gave Sal Loon the new name of Til Loon? Michael Patrick Hearn tells me that Baum knew a Van Loon family when he lived in Aberdeen, South Dakota. Perhaps one of the Van Loons was named Tillie.

Jack Snow had a plausible conjecture about "Woot." Move the "T" from back to front to form TWOO, the initials of the title of the book in which he appears! Woot says he grew up in a small town, where the inhabitants were stupid, in a "top corner" of the Gilliken Country near Oogaboo. Oogaboo is a small, poor kingdom, as we know from the first chapter of *Tik-Tok of Oz*, occupying a valley in the purple mountains at the northwest corner of Winkie Country, just across the Gilliken/Winkie border from Woot's hometown.

There has been much speculation about Nimmie Amee's name. Scholars have noticed that Amee resembles the Latin word for love, *amour*, and Nimmie is Latin for "too much." This has prompted one Oz fan (see *The Baum Bugle*, Autumn 1974) to interpret her name as meaning a girl who is loved or who loves too much. My guess is that Baum switched the "M" and "Ns" of Minnie to get Nimmie, and added a different spelling of Amy for the girl's last name.

The meanings of other names are obvious. Nick Chopper is a wood chopper. Captain Fyter is a fierce fighter. The Hip-po-gy-raf has the body of a hippo and the neck of a giraffe. Chopfyte combines Chopper and Fyter. Grunter Swyne is a grunting swine, and his wife Squelina is a squealing swine. Tommy Kwikstep makes quick steps with his twenty legs.

A major difficulty concerns the history of the nine tiny piglets. In *Dorothy and the Wizard in Oz*, the Wizard carries them in his pockets to use for performing conjuring tricks. He tells Dorothy that a sailor brought them to Los Angeles from the Island of Teenty-Weent, and he gave the man nine circus tickets for them. *The Tin Woodman of Oz* has a different story. The travelers visit the home of Professor Grunter Swyne and his wife. The Professor of Cabbage Culture and Corn Perfection claims that the nine tiny piglets are their children. The Wizard visited the Swyne home, he says, and carried off the piglets to give them an education in the Emerald City.

Michael Patrick Hearn called attention to these conflicting histories in *The Baum Bugle* (Christmas 1969), and Fred Meyer discussed possible resolutions in the Spring 1970 issue. I share Hearn's opinion that the Wizard, who was a con artist before he reformed, simply lied about the sailor and a mythical island. He must have acquired the piglets from their parents before he left Oz in his balloon, taking the little pigs with him. When he professed astonishment at hearing them talk in *Dorothy and the Wizard in Oz*, he either pretended to be amazed or he was surprised they could talk outside of Oz.

The Wizard trained the piglets to perform on their own. In the last chapter of *The Road to Oz*, eight of them entertain dinner guests with amusing antics. The ninth piglet is missing because, as we learn in chapter 17 of *Dorothy and the Wizard in Oz*, the Wizard gave it to Ozma for a pet. Later, at Ozma's birthday banquet, all nine leap out of the pie to dance on the table.

Mrs. Yoop, the giant Yookoohoo witch, provides much comic action in the quest for Nimmie Amee. We had met her giant husband in *The Patchwork Girl of Oz* where he is confined to a Quadling cave as the "largest untamed giant in captivity." He is twenty-one feet tall, weighs 1,640 pounds, is four hundred years old, and likes to eat meat people, animals, and orange marmalade. The giant greets Dorothy with "Yoho," an expression that may have suggested "Yookoohoo" to Baum, who was fond of double-o words.

After Ozma turns Mrs. Yoop permanently into a Green Monkey who no longer has magic powers, the only remaining Yookoohoo witch in Oz is Reera the Red in Baum's last Oz book, *Glinda of Oz*. When the Green Monkey's form was transferred from Woot to Mrs. Yoop, did it change from male to female?

Jinjur, the attractive feminist who led the women's Army of Revolt in *The Marvelous Land of Oz*, is visited by Ozma and Dorothy on their way to meet the Tin Woodman's group to restore them to their original forms. We learn that Jinjur's paintings are so realistic that she once painted fresh straw so the Scarecrow could take it from the canvas and use it to stuff himself. Jinjur had married a dairy farmer, but apparently they are separated or divorced because he is nowhere to be seen. Jinjur now raises candy on her ranch.

When Ozma's magic restores the Tin Woodman, Woot, the Scarecrow, and Polychrome to their natural forms, note that her sorcery obeys mysterious laws of science. In restoring Polychrome, for instance, it is necessary to go through the intermediate bodies of a dove, a speckled hen, a rabbit, and a fawn. Similarly transitional forms were required when Glinda restored Bilbil the goat to Prince Bobo in *Rinkitink in Oz*. First Bilbil became a lamb, then an ostrich, then a Tottenhot, then a Mifkit, and finally the Prince.

As in so many of Baum's Oz books, *The Tin Woodman of Oz* contains amusing political satire. Bal is made king of Loonville because he lacks common sense. He is a balloon who floats high above his throne, fastened to it by a long string. Baum surely had in mind that kings today play almost no role in a nation's affairs. They are pulled down from their heights of social standing to be seen only on ceremonial occasions. The Tin Woodman admits that his title of emperor means nothing because the people really rule themselves, and he is a mere figurehead. The golden rule that allows Ozians to get along so well with one another is the familiar parental admonition "Behave Yourself." The command is as vacuous as a politician's declaration that we must do what is right, or the minister's assertion that evil is a bad thing.

It is Baum's constant attention to small details that gives verisimilitude to what otherwise would be preposterous fantasy. Like Holmes and Watson, the Tin Woodman and the Scarecrow are such old friends that they often have nothing to say to each other.[3] The tin man oils his throat when he is hoarse. A sleeping Woot uses the Scarecrow as a pillow. After Nick Chopper's leg is cut off, he has to hop to Ku-Klip on his other leg. When the other leg is cut off, he hops to Ku-Klip on his tin leg. A loon tries to puncture the Scarecrow with a sharp thorn, but of course it has no effect. When he tries to explode the Tin Woodman, the thorn is blunted, but sharp enough to cause Woot pain. Mrs. Yoop's laugh produces a breeze so strong that it almost blows over the Scarecrow. When she walks, the stone floor of her castle trembles.

As always, Baum's frequent use of colors adds vividness to scenes in a child's mind. There are, of course, the dominant colors of the five Oz regions. Colors also are often assigned to animals and things. In *The Tin Woodman of Oz*, we have a green monkey, a brown bear, a blue rabbit, and the familiar red wagon drawn by the Sawhorse.

Douglas Greene (in *The Baum Bugle*, Autumn 1976) disclosed that the first four lines of Captain Fyter's song appear word for word as the first stanza in Baum's earlier work *Father Goose: His Book*. Here are the poem's second and third stanzas which Baum replaced with four new lines:

And when he fires his musket off
He loads it up again;
And when he charges on the foe
Resistance is in vain.

And when he marches home again
He's called a hero bold,
And many various wondrous tales
Are by the soldier told.

[3] In the second paragraph of "The Yellow Face," in *Memoirs of Sherlock Holmes*, Watson describes a walk with Holmes through a park in early spring. "For two hours we rambled about together, in silence for the most part, as befits two men who know each other intimately."

The Scarecrow's five-stanza recitation strikes me as one of the better poems in Baum's Oz books. Is it because it was not written by Baum but by Professor Wogglebug?

Some new facts about Oz come to light in *The Tin Woodman of Oz*. Winkie tinsmiths—Ku-Klip is not the only magic tinsmith in the region—make wind-up tin birds that flap about and chirp like tin whistles. We learn that Oz was once an ordinary country that became a fairyland when it was enchanted by Queen Lurline, a powerful fairy. No one in Oz ever grows older, including babies in cribs. No one gets sick. Although one cannot die of illness or old age, there are rare occasions when they can be totally destroyed. The Wicked Witch of the East, for example, turned to dust when Dorothy's house fell on her. The Wicked Witch of the West dissolved when Dorothy threw a pail of water over her.

Ozma, we are told, has decreed that no magic can be practiced in Oz except by herself, Glinda, and the Wizard. Polychrome, the Rainbow's dancing daughter, is exempt from this order because she is not an Oz resident. She visits Oz only when a rainbow follows a downpour. In this book, she uses her magic powers to restore Tommy Kwikstep to his former shape, at the same time banishing the corns on his ten remaining toes.

The book ends with Ozma sending Captain Fyter off to keep peace in the rowdy Gilliken Country where he will not be confused with his tin twin.[4] Woot is allowed to wander over Oz as he pleases. I'm surprised that no subsequent Oz book has reported on the lad's later adventures.

Some final words for collectors. *The Tin Woodman of Oz* was the last Oz book published by Reilly & Britton. The firm changed its name the following year to Reilly & Lee. Only one printing of the book bears a Reilly & Britton imprint. Its red cloth edition has twelve full-color plates by John R. Neill. The cover's paper label shows Woot sitting on the arms of the two tin men and holding their noses. Neill's black-and-

[4] Fred Meyer called my attention to Robert R. Patrick's short story "The Tin Woodman and the Tin Soldier," first published in *The Baum Bugle*, Christmas 1963. It tells how the two tin men together eliminate a field of fierce wild flowers in the Gilliken Country's great purple forest. The Tin Soldier, having forgotten his first name, is given the real name of Abel by the Tin Woodman because he is such an "able fighter."

white endpapers depict thirteen Ozians seated at a large circular table. As Dick Martin was the first to notice, the pretty woman third from left, with the initials M.C. on her blouse, is the actress Margaret Carroll, who married Neill the year after the book was published. In 1952, Little Golden Books issued an abridgement of *The Tin Woodman of Oz.* Its cover shows the tin man smoking a pipe while seated on his throne! Because he has no lungs, we must assume he doesn't inhale.

In *The Baum Bugle* (Spring 1972), David Greene quotes from a letter that Frank Reilly sent to Baum on August 10, 1916:

> As to your idea of a book for 1918 entitled The Twin Tin Woodmen of Oz; After giving this matter considerable thought, I have come to the conclusion that your theme ought to work out in great shape, but that the title is poor. Why not stick to the theme but simply call the book The Tin Woodman of Oz? The old character can, of course, be the dominating one in the book. You will probably have to give him a name as well as invent one for his twin brother. All of which is respectfully submitted

Is it possible Baum originally intended the two tin men to be blood brothers? It seems more likely that Reilly assumed this mistakenly.

Baum's handwritten first draft of *The Tin Woodman of Oz* is owned by the University of Texas, at Austin. It omits all mention of Tommy Kwikstep, the Hip-po-gy-raf, and the visit to the Swynes. I am told that the original manuscript of the Swyne chapter is owned by Yale University, and the original manuscript of the chapter about Tommy Kwikstep is in a private collection.

Hundreds of textual changes of Baum's first draft were made later either by Baum or by editors. There is no mention of Yookoohoos in the first draft. Instead, Mrs. Yoop is called a "Whisp" a word never used by Baum before or later. After Woot punctures Til Loon, we are told that another "gal Loon" met the same fate—a pun on "galloon," a trimming of lace embroidery. This sentence was removed.

Corrections were made in the Scarecrow's recitation. The first stanza's last line originally was "To be stuffed into it wherever I go." Instead of "thrills" in the second stanza, the word is "shakes." "Wheatstraw" is "oat-straw" in the fourth stanza, and "mussed-up, or dusty" in the last stanza is "crumpled and dusty."

For the next to last line of the Tin Soldier's poem, Baum added to his first draft the following alternate lines:

He likes to fight both day and night,
He likes to fight when in the right,
He takes delight in manly fight . . .

Baum dedicated *The Tin Woodman of Oz* to Frank Alden Baum, the son of L. Frank Baum's son Frank Joslyn Baum, who was author of *The Laughing Dragon of Oz* and co-author with Russell MacFall on their Baum biography, *To Please a Child.* Alden, as he was called, became a building contractor who lived in Beverly Hills. He and his wife, Louise, had no children.

Reilly & Lee discontinued the book's color plates in 1935. In 1955 they issued a new edition with pictures by Dale Ulrey. Sales were so poor that the firm abandoned plans for "modernizing" Oz art, and happily went back to Neill's immortal illustrations.

PART V

Moonshine

22

Little Red Riding Hood

> Little Red Riding Hood
> Went walking through a wood.
> She met a wolf and stopped to chat.
> Don't ask what happened after that!
>
> —Armand T. Ringer

One of the funniest of all games played by Freudian literary critics is that of finding sex symbols in old fairy tales. It is a very easy game to play. Freud is said to have once remarked that a cigar sometimes is just a cigar, but psychoanalysts who write about fairy tales seem incapable of seeing them as just fantasies intended to entertain, instruct, and at times frighten young children.

Bruno Bettelheim's analysis of Little Red Riding Hood (LRRH), in his book *The Uses of Enchantment: The Meaning and Importance of Fairy Tales* (1976) is a prime example of Freudian symbol searching. But first, a brief history of this famous fable.

The story began as a folk tale that European mothers and nurses told to young children. The fable, in its many variants, came to the attention of Charles Perrault (1628–1703), a French attorney turned poet, writer, and anthologist. He published one version in a 1697 collection of fairy tales—a book that became a French juvenile classic.

Perrault opens his story "*Le Petit Chaperon Rouge*" (Little Red Cape) by telling about a pretty village girl who is called Little Red Riding

Hood because she loves to wear a red cape and hood given to her by her grandmother. Her mother hands her some biscuits and butter to take to the sick grandmother in a nearby village. Walking through a wood, LRRH encounters a friendly wolf who asks where she is going. After she tells him, the wolf says he'll go there too, but by a different route and they'll see who gets there first.

The wolf arrives ahead of the girl, devours the grandmother, then crawls into bed. When LRRH shows up he simulates the grandmother's voice, telling her to put the biscuits and butter aside and climb in bed. LRRH undresses and does as she is told. A famous dialog follows: "What great arms you have, grandma! The better to embrace you, my child. What great legs you have! The better to run with, my child. What great ears! The better to hear with. What great eyes! The better to see with. What great teeth! The better to eat you with."

The wolf then gobbles up LRRH and the story ends! I have been told, though I strongly doubt it, that French children find this ending amusing, and are not in the least disturbed by it. Andrew Lang, who reprinted Perrault's version in his *Blue Fairy Book*, severely criticizes Perrault for choosing a version with such a gruesome ending.

When the German brothers Jacob and Wilhelm Grimm later published in 1812 their collection of more than 200 traditional fairy tales, many taken from Perrault, they gave the story a less grim ending. In their version (you'll find it in the Modern Library's *Tales of Grimm and Andersen*), LRRH's mother gives her cake and a bottle of wine to take to the ailing grandmother. LRRH is not afraid of the wolf when she meets him in the forest. He persuades her to pick some flowers to take to her grandmother. While she is doing this (disobeying her mother who told her not to dawdle) the wolf hastens to the grandmother's house, finds the door unlocked, enters, and promptly eats the grandmother.

When LRRH arrives she is surprised to find the door open. She thinks it is her grandmother in bed because the wolf has pulled a nightcap over his face, and sheets over his body. LRRH stands beside the bed while the familiar dialog occurs about the wolf's body parts. The wolf then springs out of bed and eats LRRH. He now goes back to bed and falls asleep. A passing hunter hears the wolf's loud snores. He goes inside to investigate and is about to shoot the wolf until he realizes it may have eaten the grandmother. So he pulls out a knife and cuts open the wolf's belly. Both LRRH and the grandmother emerge as unharmed as Jonah when he was vomited out of the whale's belly.

LRRH brings some big stones into the house to put inside the wolf, who is still asleep. When he awakes and tries to get away, the heavy stones drag him down and he drops dead. The hunter skins the wolf and takes the skin home. The grandmother can hardly breathe, but she feels much better after eating the cake and drinking some wine. LRRH says to herself, "I will never again wander off into the forest as long as I live, when my mother forbids it."

The tale is short and simple. Its obvious moral is that children should obey their mothers when they walk through dangerous areas, and to beware of seemingly friendly strangers. I suppose it is the linking of LRRH's beauty and innocence with her grisly experience that has led to her capturing the hearts of so many adults everywhere, especially in Germany, France, Sweden, and England. "Little Red Riding Hood was my first love," declared Charles Dickens. "I felt that if I could have married Little Red Riding Hood, I should have known perfect bliss."

Bruno Bettelheim devotes eighteen pages of his book on fairy tales to LRRH.[1] In his eyes the girl is not as innocent as she seems. She is at the nymphet stage when her premature "budding sexuality" is creating deep unconscious conflicts between her id (animal nature) and her superego (conscience), as well as between her allegiance to what Freud called the "pleasure principle" and the "reality principle." Unconsciously, she wants to be seduced by her father. The wolf's eating her represents that seduction.

The red color of LRRH's hood, according to Bettelheim, symbolizes her unconscious sexual desires. He sees the gift of the hood by the grandmother as representing a transfer of sexual attractiveness from an

[1] Alan Dundes, an anthropologist at the University of California, Berkeley, in an article in *The Journal of American Folklore*, accused Bettelheim of shamelessly cribbing from *A Psychiatric Study of Fairy Tales* (1963), by Stanford psychiatrist Julius Hauscher. Not only does Bettelheim take up the tales in the same order, but there is what Dundes calls "wholesale borrowing of key ideas." Nowhere in his book or in a long list of references does Bettelheim mention Hauscher. See "Was He Really Borrowheim?" in *Newsweek*, February 18, 1991, p. 75.

Little Red Riding Hood and the Wolf (Gustave Doré)

old sick woman to a young healthy girl. The grandmother is a symbol of the little girl's mother. When the wolf eats the grandmother it represents the little girl's wish to get rid of her mother so she can have her father all to herself.

In Grimm's version, Bettelheim sees the hunter as another father symbol. When he cuts open the wolf's belly it indicates "the idea of pregnancy and birth," thus coming "too close for comfort in suggesting a father in a sexual activity connected with his daughter."

Bettelheim, of course, is not the only Freudian to read dark sexual meanings into the story. Psychoanalyst Erich Fromm, in *The Forgotten Language: An Introduction to the Understanding of Dreams, Fairy Tales, and Myths* (1951) is also convinced that LRRH is experiencing unconscious sexual impulses and really wants to be seduced by the wolf. The red cape symbolizes her menstrual blood as she enters womanhood. When the mother warns her not to leave the path or she might fall and break the wine bottle, it represents the mother's fear that her daughter might lose her virginity by breaking her maidenhead.

"The male is portrayed as a ruthless and cunning animal," Fromm writes. The sexual act becomes a "cannibalistic act in which the male devours the female." Fromm sees this as an expression of a deep antagonism toward men by frigid females who do not enjoy sex. The male wolf is "made ridiculous" by showing "that he attempted to play the role of a pregnant woman, having living beings in his belly." The stones that LRRH puts in the wolf's stomach are "symbols of sterility" that cause him to collapse and die. The stones "mock his usurpation of the pregnant woman's role."

"The story," Fromm concludes, "speaks of the male-female conflict; it is a story of triumph by man-hating women, ending with their victory, exactly the opposite of the Oedipus myth, which lets the male emerge victorious from this battle."

Jack Zipes, who teaches German at the University of Minnesota, is the author of *The Brothers Grimm* (1986), a two-volume edition of the Grimms' stories, a collection of French folk tales, and other books on folklore. One of his books is titled *The Trials and Tribulations of Little Red Riding Hood* (1983, updated in 1993). The book is a marvelous scholarly history of the LRRH fable and its many versions and interpretations.

Zipes covers all the oral variations that preceded Perrault, as well as the many retellings by writers from the Grimm brothers to 1993. Some of the oral tales are even more morbid than Perrault's version. In several versions the Wolf slices up the grandmother and pours her blood into a bottle. LRRH then eats and drinks what she thinks is meat and wine before the wolf eats her. In other versions LRRH escapes by telling the wolf she has to go outside to relieve herself.

Thirty-eight variations of the tale are reprinted in Zipes's anthology, along with a raft of illustrations from books and advertisements. At the back of the book he lists 147 published versions of the story, including retellings by Walter de la Mare and James Thurber, as well as comic parodies, poems, plays, recordings, musicals, and films. His bibliography of critical references runs to 153 items!

Zipes takes both Fromm and Bettelheim to task for not recognizing the story's male bias, namely the view that girls secretly want to be raped, and may even encourage it, and that they need a good strong man to shield them from such desires. "It is because rape and violence," Zipes writes in his preface, "are at the core of the history of Little Red Riding Hood that it is the most widespread and notorious fairy tale in the Western World."

Zipes greatly admires today's feminist writers who analyze the story from a female perspective. Most of them deny that LRRH had unconscious impulses to be raped. They give her the strength and cleverness to take care of herself. An amusing example, though not by a feminist, is Thurber's burlesque in *Fables for Our Times and Famous Poems* (1939). In this brief account, when LRRH approaches the wolf she sees at once he is not her grandmother, "for even in a nightcap a wolf does not look any more like your grandmother than the Metro-Goldwyn lion looks like Calvin Coolidge. So the little girl took an automatic out of her pocket and shot the wolf dead. Moral: It is not so easy to fool little girls nowadays as it used to be."

Views similar to those of Fromm and Bettelheim are advanced in greater detail by Carl-Heinz Mallet, who runs a school for retarded children in Hamburg. A dedicated Freudian, he has written a study titled *Grimms' Fairy Tales*, and another book, *Fairy Tales and Children*. Mallet sees LRRH as a nymphet who is curious about sex. "Behind a sweet, pure facade," he writes in the second-mentioned book, "nature dominates, and a primal life seethes." He finds LRRH's sex appeal brought out more vividly in a version of the tale told by Ludwig Bechstein, another German folklorist.

"Once there was an absolutely darling, charming little slip of a girl," is how Bechstein opens his version. When the wolf sees her he thinks, "Oh, you dearest, appetizing hazelnut, you—I have to crack you." Mallet thinks the red hood is the girl's signal that she wants to be raped. "For all her innocence, Little Red Riding Hood turns men on." Why is she so friendly with the wolf? Because, writes Mallet, she subconsciously wants him to attack her.

What about the grandmother? Mallet believes she is a symbol of the mother's sublimated longing for fresh sexual experiences, symbolized, of course, by the grandmother being eaten by the wolf. Oral sex with a vengeance! Part of the mother "is worried about her daughter. Another part wants this very thing for herself."

The wine and cake, for Mallet, represent "bodily pleasures." The grandmother leaves her door unlocked because she secretly hopes a man will invade the house and rape her. Mallet puts it this way:

> The mother dares not seek out the wolf for herself, much as she might like to, and so dresses her daughter attractively and sends her out into the forest, the home of the wicked wolf. Simultaneously innocent and sexy, the child will exert a powerful attraction; the wolf will not stop at voicing polite compliments, and the mother knows it.

Of course the mother is not consciously aware of her motives. She really wants her daughter to seduce the wolf for her (the mother's) vicarious pleasures! The various parts of the wolf's body that LRRH sees and feels represent, for all Freudian critics, the one male organ that she envies and most desires. They suspect that if the truth were known, LRRH finally exclaimed, "What a big thing you have between your thighs!" LRRH's death—you guessed it—is a symbol of her orgasm as she is raped. Why is the tale so universally popular? "Not even the most graphic book or movie," Mallet assures his readers, "could outdo the sensuous scene in the grandmother's bed."

The identification of the wolf with a male human womanizer is the basis of a 1966 hit song by Sam the Sham and the Pharoahs. The music is by Ronald Blackwell. The song opens with a wolf howl followed by, "Who's that I see walking in these woods? Why, it's Little Red Riding Hood!" The song goes on to say that LRRH has big eyes of the sort that drive wolves mad, and "little big girls" should not go strolling through "spooky old woods" alone.

Sam then offers to protect LRRH by walking with her each time she goes through the forest. He hopes she will trust him. Although he yearns to hold her, he won't try because she might suppose *he's* a big bad wolf. However, even bad wolves can be good, and he has a big heart, the better to love her with. He'll be happy to accompany her on walks through the woods until she realizes he's no threat.

Wolf calls are interspersed between the song's stanzas. The recording ends with a wolf howl and Sam repeating, "I mean baaaa! baaaa! baaaa!"

Two poems by James Whitcomb Riley are worth mentioning. His three-stanza tribute, "Red Riding Hood," opens:

Sweet little myth of the nursery
story—
Earliest love of mine infantile breast . . .

Riley's other poem, "Maymie's Story of Red Riding Hood," is a lengthy ballad told in an annoying dialect. After the wolf eats the grandmother, a wood-chopper saves LRRH by killing the wolf with a blow on his head, but the poor grandmother remains unrecovered.

Let's try a non-Freudian interpretation of the fable—one that could be made by a theologian. The story is an allegory about good and evil. LRRH is the child, innocent of sin, who is slain by the irrational evil of the universe. Like Moby-Dick, the wolf represents this inescapable aspect of our lives. Every year millions of beautiful children are either slain by a dreadful disease or by slow starvation. In the end, we are all killed by the big bad wolf. In the German Protestant version, the hunter is God. He slays the monster Satan, and restores LRRH and her grandmother to everlasting life.[2]

But enough of such baloney. My next chapter will recount the sad life of Bruno Bettelheim, how his views about autistic children did terrible harm both to the children he treated and their parents, and about his eventual suicide.

[2] On Christian interpretations of five of the best-known fairy tales by Grimm, including LRRH, see *The Owl, the Raven, and the Dove: The Religious Meaning of the Grimms' Magic Fairy Tales*, by G. Ronald Murphy (2000).

Little Red Riding Hood
Guy Wetmore Carryl

Most worthy of praise were the virtuous ways
Of Little Red Riding Hood's ma,
And no one was ever more cautious and clever
Than Little Red Riding Hood's pa.
They never misled, for they meant what they said,
And frequently said what they meant.
They were careful to show her the way she should go,
And the way that they showed her she went.
For obedience she was effusively thanked,
And for anything else she was carefully spanked.

It thus isn't strange that Red Riding Hood's range
Of virtues so steadily grew,
That soon she won prizes of various sizes,
And golden encomiums too.

As a general rule she was head of her school,
And at six was so notably smart
That they gave her a check for reciting "The Wreck
Of the Hesperus" wholly by heart.
And you all will applaud her the more, I am sure,
When I add that the money she gave to the poor.

At eleven this lass had a Sunday-school class,
At twelve wrote a volume of verse,
At fourteen was yearning for glory, and learning
To be a professional nurse.
To a glorious height the young paragon might
Have climbed, if not nipped in the bud,
But the following year struck her smiling career
With a dull and a sickening thud!

(I have shed a great tear at the thought of her pain,
And must copy my manuscript over again!)

Not dreaming of harm, one day on her arm
A basket she hung. It was filled
With drinks made of spices, and jellies, and ices,
And chicken wings, carefully grilled,
And a savory stew, and a novel or two
She persuaded a neighbor to loan,
And a Japanese fan, and a hot-water can,
And a bottle of eau de cologne,
And the rest of the things that your family fill
Your room with whenever you chance to be ill.

She expected to find her decrepit but kind
Old grandmother waiting her call,
Exceedingly ill. Oh, that face on the pillow
Did not look familiar at all!
With a whitening cheek she started to speak,
But her peril she instantly saw:
Her grandma had fled and she'd tackled instead
Four merciless paws and a maw!
When the neighbors came running the wolf to subdue,
He was licking his chops—and Red Riding Hood's, too!

At this terrible tale some readers will pale,
And others with horror grow dumb,
And yet it was better, I fear, he should get her—
Just think what she might have become!
For an infant so keen might in future have been
A woman of awful renown,

Who carried on fights for her feminine rights,
As the Mayor of an Arkansas town,
Or she might have continued the sins of her 'teens
And come to write verse for the Big Magazines!

THE MORAL: There's nothing much glummer
Than children whose talents appall.
One much prefers those that are dumber.
And as for the paragons small—
If a swallow cannot make a summer,
It can bring on a summary fall!

23

The Brutality of Dr. Bettelheim

Dr. Bruno Bettelheim
Devoted most of his time
Condemning mothers to perdition
By blaming them for their autistic child's condition.

Friend Armand T. Ringer composed the above clerihew after reading a first draft of this column. It is an accurate statement about the enormous harm that can be done by dogmatic, closed-minded Freudians.

Estimates of the number of children afflicted with a broadly defined autism vary from one in a thousand to one in 250. It is more common than Down syndrome. Boys outnumber girls four to one. Symptoms start to appear when a child approaches two, but often are not recognized as autistic until the child begins school. Autistic children look deceptively normal, and many are very beautiful.

Like almost all mental illnesses, autism has a spectrum of symptoms that range from mild to severe. Severe autism has the following traits:

1. Children with autism are self-absorbed ("auto" is Greek for "self"). They live inside a glass shell, hardly recognizing the existence of their parents or others. They are unresponsive to affection and incapable of normal social relationships.

2. They seldom make eye contact, and rarely point to anything. Many refuse to talk or they speak only in brief sentences. Some speak fluently

but in repetitive sentences. They are unable to maintain a normal conversation.

3. About 80 percent are mentally retarded.

4. Their behavior is bizarre. They are prone to a wide variety of repetitive rituals such as banging their head, flapping their arms, slapping their face, pulling their hair, or rocking back and forth. They can spend hours repeating such mindless tasks as running sand through their fingers in a sandbox, typing a single key on a typewriter, or spinning objects on the floor or in their hands. The slightest change in their experience, such as a mother wearing a different dress, or a toy moved to a different spot, can trigger a severe tantrum.

5. Some autistic adults resemble idiot savants in developing curious skills such as memorizing a phone book, an ability to multiply large numbers or identify huge prime numbers, or quickly memorize complicated musical scores. Some become obsessed with calendars and can correctly name the day of the week for any given date. Some rapidly solve jigsaw puzzles even when the pieces are picture-side down. Such strange abilities, along with the usual self-absorption, were featured in the movie *Rain Man*, a film in which Dustin Hoffman takes the role of an autistic middle-aged man.[1]

"Joey: A Mechanical Boy," by Dr. Bruno Bettelheim (*Scientific American*, March 1959), is a famous article about an autistic child who thought he was a robot. He would construct all sorts of weird machines around his bed, and repeatedly connect himself to them to obtain power for running himself. It would be interesting to know Joey's later history.

Autism is a mysterious malady whose causes are not known. There are several competing theories, none confirmed, but all recognize that autism is a brain disorder probably caused by a set of malfunctioning genes. If one of a pair of identical twins is autistic, the other twin will be autistic about 65 percent of the time. The most persistent myth about autism is that there is a normal child inside the shell desperately wanting

[1] Such abilities can be greatly exaggerated. For example, a *Newsweek* cover story "Understanding Autism" (July 31, 2000), cited an autistic savant who could memorize a phone book in ten minutes. No one can even *read* a phone book in ten minutes, let alone memorize it.

Dr. Bruno Bettelheim (Corbis)

to emerge if only the shell can be shattered. There is no normal child inside the shell.

For a report on recent research into the causes of autism, see "The Early Origins of Autism," by Patricia Rodier, professor of obstetrics at the University of Rochester, in *Scientific American*, February 2000, pages 56–63. See also Uta Frith's earlier article "Autism," in the same magazine, June 1993. A London psychologist, Frith is also the author of *Autism: Explaining the Enigma* (1989).

Evidence that autism is in any way related to how parents behave is unconvincing, nor is there evidence that it is related, as so many parents foolishly believe, to early vaccinations. Marian DeMeyer, an Indiana University psychiatrist, made a careful study of three groups: parents with an autistic child, parents with normal children, and parents with a brain-damaged child. Personality tests showed that the three groups were indistinguishable. (See DeMeyer's paper in *The Journal of Autism and Childhood Schizophrenia*, Vol. 2, 1972, pages 49–66.) More recent work, I am told, has cast doubt on DeMeyer's findings.

Siblings of autistic children are normal. The percentage of children with autism is the same in all cultures, and among all racial, ethnic, and socioeconomic groups. Medication and therapy help mildly autistic children lead normal lives, but for severe autism no cures are known.

Strong evidence that autism is a dysfunction of the brain has been available for half a century, and was taken for granted by neurologists outside the Freudian tradition. For a while it was called childhood schizophrenia. However, psychoanalysts and amateur Freudians persisted for decades in the fantasy that autism was somehow caused by unloving parents, especially by cold "refrigerator mothers." The leading advocate of this absurd view was Dr. Bruno Bettelheim (1903–1990).

Bettelheim was a small, bald, nearsighted man with thick glasses and a strong Austrian accent. He liked to tell people he was so ugly that when his mother first saw him after his birth she exclaimed, "Thank God it's a boy!" Born in Vienna to a Jewish father who died of syphilis, Bettelheim claimed to have studied under Freud. Although he was briefly psychoanalyzed in Vienna, he was not trained as an analyst. He never claimed to be a psychiatrist of any sort—only a psychologist. His doctorate in Austria was on the aesthetics of nature.

Arrested by the Nazis, Bettelheim spent a year in Nazi concentration camps, first at Dachau, then at Buchenwald. This was before they became extermination centers. Released in 1939, he came to the United States where he was given a job teaching at Rockford College for women, near Chicago, and later at the University of Chicago. In 1944 he took over the University of Chicago's decaying Sonia Shankman Orthogenic School for disturbed children, which he headed from 1944 to 1978. During this period he became one of the country's most respected experts on childhood pathology.

Bettelheim wrote eleven books, numerous articles in technical and popular journals, and an advice column that ran for ten years in *The Ladies Home Journal.* He lectured everywhere, and even appeared as a psychiatrist in Woody Allen's 1983 film *Zelig.*

In 1983 Bettelheim retired from the Orthogenic School, or Bruno's Castle as it was sometimes called, and moved to California. He had early on been divorced from Gina, his first wife. His second wife, Gertrude, died in 1984 after forty-three years of a happy marriage and three children. After a stroke in 1987, Bettelheim moved to a retirement home, which he despised, in Silver Spring, Maryland. In 1990, ill, lonely, and depressed over a rift with his daughter Ruth, he overdosed on sleeping pills, fastened a plastic bag over his head, and died.

After his suicide, evidence of Bettleheim's dark side began to emerge. Although many of his counselors at the Orthogenic School considered him brilliant and admirable, others began openly to call him a cruel, egotistical tyrant, the guru of a cult, and a power-mad mountebank.

Evidence accumulated that Bettelheim exaggerated his bravery in the concentration camps, and lied when he said that Eleanor Roosevelt had helped him escape. He claimed an 85 percent cure rate of autistic children in his care, a boast no other psychiatrist came close to making. Critics pointed out that Bettelheim alone diagnosed autism and he alone evaluated "cures." Most of his cures, they charged, were of children who were not even autistic or only mildly so.

Although untrained in analysis, Bettelheim was a Freudian fundamentalist. Counselors reported that every trivial incident that occurred in his school, such as a child breaking a dish or unintentionally hitting another child with a rubber ball, was taken by Bettelheim to be an unconscious expression of hostility. He was given to outbursts of anger and frequently slapped children. Alida Jatich, a patient for seven years who became a computer programmer in Chicago, published an article in *The Chicago Reader* (a weekly newspaper) in which she said Bettelheim once dragged her naked and dripping from a shower and slapped her repeatedly in front of her dorm mates. In her words:

> In person, he was an evil man who set up his school as a private empire and himself as a demigod or cult leader. He bullied, awed, and terrorized the children at his school, their parents, school staff members, his graduate students and anyone else who came into contact with him.

Roger Angres, another former patient, in an article for *Commentary* (October 1990), described what he called Bettelheim's "insulting and intimidating theatrics." He insulted children, Angres wrote, "just in order to break any self-confidence they might have. I lived in terror of his beatings, in terror of his footsteps in the dorm."

The degree of Bettelheim's cruelty toward patients was mild compared to his cruelty toward mothers. For a detailed account of this I recommend science writer Edward Dolnick's excellent, hard-hitting *Madness on the Couch: Blaming the Victim in the Heyday of Psychoanalysis* (Simon and Schuster, 1998). It is one of a raft of recent books exposing psychoanalysis as one of the most monumental pseudosciences of the last century. What follows is taken mainly from Dolnick's book.

Bettelheim was convinced, in spite of overwhelming evidence to the contrary, that autism had no organic basis but was caused entirely by cold mothers and absent fathers. "All my life," he wrote, "I have been working with children whose lives have been destroyed because their mothers hated them."

Again: "The precipitating factor in infantile autism is the parent's wish that his child should not exist."

In the mid-fifties Bettelheim adopted a policy known as "parentectomy." Under this policy, parents were not allowed to see their children for at least nine months!

You can imagine the desolation felt by mothers when they were told they had created their child's pathology. Annabel Stehli, one of many such mothers, read Bettelheim's book about autism, *The Empty Fortress* (1967), a book his detractors called The Empty Book. In *The Sound of a Miracle* (1991), Stehli described her reaction this way:

> I was carrying around this terrible secret. I didn't want to talk to anyone about Bettelheim. My husband said that he thought it was baloney, but I didn't talk to my friends about it. I was very alone. I really felt as if I had a scarlet letter on, only the "A" was for "Abuse."
>
> I felt that I'd hurt Georgie in some subtle way that I couldn't grasp, and if I could just figure it out, then maybe she'd be okay. There was a part of me that *wanted* to believe Bettelheim, because that would mean that if I got better, Georgie would get better.

A typical bit of Freudian nonsense in Bettelheim's *Empty Fortress* was how he explained a child's obsession with weather. The child broke down the word "weather," unconsciously of course, into "we/eat/her." It's hard to believe, but Bettelheim actually wrote: "Convinced that her mother (and later all of us) intended to devour her, she felt it imperative to pay minutest attention to this 'we/eat/her.'"

Other Freudian analysts, as well as scientists who were not psychiatrists, followed Bettelheim in blaming the poor mothers for their child's autism. Psychologist Harry Harlow's research with monkeys who were deprived of mothers convinced him that autism was caused by icebox mothers. Dutch zoologist Nikolaas Tinbergen, who for his study of bird behavior shared a 1973 Nobel Prize in medicine with zoologist Konrad Lorentz and Karl von Frisch, also fell for the preposterous cold mother theory. He actually wrote a book titled *Autistic Children: New Hope for a Cure* (1983). And what was the cure? It was constant hugging of children by their mothers!

Dolnick records an ironic twist to all this. Psychiatrists Maurice Green and David Schecter, in several technical papers, argue that autism is caused not by cold mothers, but by mothers who love a child too much! They give a milk bottle before a child asks for it, or a toy before he or she wants it. By excessive anticipation of their child's needs, the child naturally doesn't bother to speak and remains a permanent infant.

There are two biographies of Bettelheim. Richard Pollak, former literary editor of *The Nation*, published *The Creation of Dr. B* in 1997. It's a bitter attack on Bettelheim, who was called Dr. B by his patients and counselors. Nina Sutton's *Bettelheim: A Life and a Legacy* (1996), translated from the Greek, is more balanced. She sees Bettelheim as a complex man with both good and bad qualities. A hostile account of a character called "Dr. V" is in Tom Wallace Lyons's autobiographical novel *The Pelican and After* (1983). Lyons was a former patient of Bettelheim.

Bernard Rimland, a psychologist and father of his autistic son Mark, is the author of *Infantile Autism* (1964), another slashing attack on Bettelheim. Dolnick devotes several pages to Mark, an amiable man who has managed to adjust to life. Mark is one of those autistics who can name the day of the week for any date, but is unable to explain how he knows. When Bettelheim died, Rimland's comment reflected the opinions of almost all authorities on autism. "He will not be missed."

My next chapter will be about Facilitated Communication, a crazy development involving autistic children that is almost as sad and deplorable as Bettelheim's attacks on refrigerator mothers.

24

Facilitated Communication: A Cruel Farce

My previous chapter was about the myth, so vigorously promoted by Dr. Bruno Bettelheim, that autism is caused by "refrigerator mothers"—mothers with cold, unloving personalities. During the past two decades other preposterous myths about autism have flourished. One is the widespread belief among agonized parents that their child's autism is caused by early vaccinations. Because there is not a shred of evidence for this, I shall not waste space on it here. My topic is a much more pervasive, more cruel myth—the belief that hiding inside the head of every child with autism, no matter how severe, is a normal child whose intelligent thoughts can emerge through a curious technique called facilitated communication (FC).

Here's how FC works. An autistic child is seated at a typewriter or computer keyboard, or perhaps just a sheet of paper with the keyboard drawn on it. (Although autism strikes both sexes, males outnumber females four to one so I shall use the pronoun "he" for any child with severe autism.) A therapist, usually a woman, is called the child's "facilitator." She asks the child a question, then grasps his hand, wrist, or elbow—usually the hand—while the child extends his index finger and begins to type. The belief is that the child has the ability to communicate intelligent thoughts by typing, but lacks the muscular coordination needed for finding the right keys. The facilitator assists him in locating the keys she is sure he intends to hit.

A wondrous miracle now seems to take place. Although the child has been thought to be mentally retarded, unable to read, write, or speak coherently, he types out lucid, sophisticated messages that could only come from a normal, intelligent child.

This amazing technique was discovered in the early 1970s by Rosemary Crossley, of Melbourne, Australia. She and her associates founded the Dignity Through Education and Language Communication Center in Melbourne. Crossley is the author of numerous papers in technical journals, and a book titled *Annie's Coming Out* about one of her patients.

Enter Douglas Biklen, a professor of education at Syracuse University. When he first visited Crossley in the late 1980s he was skeptical of her methods, but soon became convinced that she had hit upon a revolutionary new technique. Today he considers her his "dear friend" and mentor. Back at Syracuse, Biklen founded the Facilitated Communication Institute, and quickly became the nation's top guru in promoting FC. Thousands of therapists have been trained in FC at his institute.

FC spread like wildfire across the nation and abroad. Dozens of FC centers were established around the U.S. where parents, for a sizeable fee, could bring a child with autism and have his normal mind released from the prison of his terrible disorder.

You can imagine the excitement and euphoria of parents long desperate for rational conversation with a loved child. They often wept with joy when they became convinced that their child could say to them through typing, "I love you, Mom" or "I love you, Dad." Not only that, but they typed long meaningful sentences, using words they were incapable of speaking. Some even wrote poetry.

This seemingly miraculous breakthrough was enthusiastically endorsed by the print and electronic media. Articles praising FC appeared in newspapers and popular magazines. *Reader's Digest* (March

1993) published "The Secret Life of Arthur Wold," a moving account of a child released by FC from the glass shell believed to have imprisoned his mind. Movies and television shows lauded the new technique.

That something was terribly wrong with FC was obvious from the start to almost all psychiatrists, neuroscientists, and psychologists familiar with autism. For one thing, while the poor child was typing brilliant messages, he almost never looked at the keyboard. He would glance around the room, often smiling or giggling as if having fun, sometimes screaming, sometimes closing his eyes.

Skeptics checked with expert typists. Not one was capable of typing a sentence with one finger unless he or she could see the keyboard. This made no impression on facilitators or parents eager to believe. Therapists made the absurd claim that their children had the wonderful ability to memorize the keyboard so completely that they did not need to see it when they typed! Children with severe autism, unable to read, write, or speak clearly, found themselves in schools where they passed examinations in literature, arithmetic, and other subjects provided, of course, a facilitator was beside them, guiding their hand!

It quickly became obvious to everyone not caught up in the FC fad that the facilitators, although totally sincere, were naive and poorly trained. They were unaware of the strength of what is called the Ouija (or ideomotor) effect. Unconsciously they were guiding a child's finger to keys they imagined the child was seeking. In brief, it was they, not their patients, who did the typing.

As one would expect, because guiding a finger to a key is seldom accurate, the children's messages swarm with typos. Here are some typical examples reported by Biklen:

> I AMN NOT A UTISTIVC ON THJE TYP.
>
> MY MOTHER FEELS IM STUPID BECAUSE IH CANT USE MY VOICE PROPERLY.
>
> I AM VERY UPSET BECAUSE I NEED FACILITASION. I DONT WANT TO DEPEND ON PEOPLE.

I AI DONT WANT TO BE AUTISTIC. NOBODY REALLY ZUNDERSTANDS WHAT IT FEELS LIKE. IT IS VERY LONELY AND I FEEL LOUSY. MY MOOD IS BAD A LOT. I FEEL LESS LONELI WHEN I AM WITH KIDS.

Is it possible to prove beyond any doubt that facilitators do the typing? Indeed, it is so ridiculously easy that promoters of FC must be simple-minded not to have thought of such simple tests. In one early test, made by skeptics, the child and his facilitator wore headphones. When a question was heard by both child and facilitator, the child typed a reasonable response. But when only the child heard the question, while the facilitator heard only music, the child's answer had no bearing on the question.

A simple visual test was even more dramatic. A picture on a cardboard folder was shown to both child and facilitator. The child accurately typed the target's name when his hand was held. In a repeat test, the experimenter showed the facilitator a picture, but this time, as he turned the folder so only the child could see the target, he secretly moved a flap that covered the first picture and exposed a different one. You can guess what happened. The child did not type the name of the picture he saw. He typed the name of the picture that only the facilitator had seen.

This definitive, unimpeachable test was featured in a marvelous documentary, "Prisoners of Silence," aired by the PBS series *Frontline* in October 1993. Many facilitators were devastated by the revelation that they had been deceiving themselves for years. One female therapist, interviewed by *Frontline*, was in tears. A similar exposé was a *60 Minutes* program, "Less Than a Miracle," aired in 1994.

The Ouija effect is far more powerful than most people realize. It explains why the planchette, in response to questions, glides so smoothly over the Ouija board to spell answers which seem to come from spirits. It is the secret behind rotating dowsing rods, table tipping, and automatic writing. Entire books have been written by hands that seem to be moved by spirit controls.

You can demonstrate the Ouija effect easily by tying a ring, or some other small weight, to a piece of string. Tell a friend that if she holds this little pendulum over a man's hand it will swing back and forth in a straight line. Held over a woman's hand it swings in ovals. You'll be surprised by how often this works, provided of course a person is told how the weight will swing.

In the late 1990s FC took an ugly turn. Overzealous facilitators, oblivious to the fact that they were doing the typing, began to produce messages accusing parents of sexually molesting their child! These messages often contained graphic four-letter obscene words. It is hard to believe, but uninformed police and judges, on the basis of such accusations, actually arrested dozens of innocent and astounded fathers. Some even went to jail for months, exhausting fortunes on legal defenses before enlightened judges tossed their cases out of court.

A recent case in England, reported in London newspapers on July 12, 1999, involved a teenage boy with a mental age of two who was suffering from extreme autism and unable to speak. With a hand guided by a facilitator he had typed accusations of sexual abuse by his fifty-year-old father. The boy was made a ward of the court until the court found no evidence of abuse. The judge, Dame Elizabeth Butler-Sloss, branded FC a dangerous, unverified technique that should never be used again in any British court to support sexual abuse charges.

In 1992 a group of facilitators at the University of Wisconsin, in Madison, became convinced that some of their children were gifted with psychic powers. For example, while they typed with their hand held, their sentences often revealed what their facilitator was thinking. Suspecting ESP, the facilitators began giving their children tests for telepathy. They would show a child a picture, for example of an elephant. Another child, in a distant room, would type "elephant" when asked what picture the other child had seen. Of course both facilitators knew the target was an elephant. The Madison facility was and is headed by Anne M. Donnellan, a professor of education with forty years of experience working with autistic children. Convinced that her patients had psychic abilities, she invited magician and psi-buster James Randi to visit Madison and evaluate the ESP evidence.

Randi realized at once, even before making the trip from his home in Florida to Madison, what was going on. The facilitators, not the children, were doing the typing. Naturally it would seem that the typing child was reading their mind. The apparent ESP was explained by the fact that the facilitators were communicating with each other. In a letter to me before he left Florida, Randi predicted correctly that if he made a few simple tests in which the facilitator at the receiving end did not know the target, and no ESP took place, he would be met with strong resistance from the therapists. He would be accused of destroying ESP by his skepticism and his insulting experiments.

Randi's first test was to shuffle a set of cards with different pictures, then randomly select a card. A facilitator was asked to leave the room while Randi showed the picture to her child and to everyone else in the room. When the facilitator returned, held the child's hand, and asked him to name the picture he had seen, he typed only incorrect names. Similar tests for telepathy were also total failures. The receiving child was correct only when his facilitator knew the target.

As Randi anticipated, no one at the Madison facility welcomed his disclosures. The children too, their hands held, began to type such messages as "I don't like this man from Florida. He is upsetting my facilitator. Send him home."

"No one would like more than I to find that I am wrong," Randi wrote to me in his first letter. (At Madison he sent me daily reports of his investigations.) "It would be thrilling indeed to discover that autistic children, far from being incapable, isolated and mentally inferior . . . are actually capable of much more and only need to be released from their imprisonment."

Now that FC has been so thoroughly discredited, one would hope that FC leaders would admit they had been duped, and like the Arabs in Longfellow's "Day is Done," would fold their tents and silently steal away. Alas, true believers in a bogus science seldom change their minds or admit their mistakes. Biklen is still at Syracuse University, running his institute, and still convincing parents that hidden inside the glass shell of autism there is a bright, normal child who loves them and is eager to converse with them provided his hand is held by a facilitator. I should add that of course children with only mild autism can be taught to type rational messages without any help, and can grow up to lead normal, constructive lives.

Through high-priced seminars, and sales of videotapes and literature, it is estimated that Biklen is bringing millions of dollars annually to his university. This cannot, however, be the only reason why Kenneth

Shaw, the president of Syracuse University, has made no effort to maneuver Biklen off the campus. He not only defends Biklen on the grounds that a university must allow free opinions to be held by any faculty member, but he also has defended Biklen's competence. "[Biklen] has had the intellectual rigour," Shaw told a reporter, "one would expect of qualitative research in this area."

Biklen has had no training in psychology, psychiatry, or work with the disabled. His doctorate at Syracuse was in sociology. His books include *Unbound: How Facilitated Communication Is Challenging Traditional Views of Autism and Ability/Disability* (1993), and *Contested Words, Contested Science: Unraveling the Facilitated Communication Controversy* (1997). Both books were published by Columbia University's Teachers College Press. His institute publishes a periodical titled *Facilitator Communication Digest*.

Biklen's present position on FC is that some facilitators, inadequately trained, have unwittingly guided a child's hand, but he thinks they are a small minority. On his Web site in June 2000 he warned: "Facilitated Communication should never involve guiding a person as he or she attempts to point or type."

Donnellan edited *Classic Readings in Autism* (1985), published by Teacher's College Press. Her latest book, written in collaboration with Martha Leary, is *Movement Differences and Diversity in Autism/Mental Retardation* (1997), published by the DRI Press, Madison. Like Biklen, she grants that the Ouija effect may explain some of the typing by children with autism, but she firmly believes that the effect can be avoided by well-trained facilitators, and that most of the criticism of FC is unfounded.

In 1994 the American Psychological Association and the American Speech-Language-Hearing Association each declared that FC is without scientific validation.

In recent years the belief that therapists, using hypnosis and other techniques, could revive memories of childhood sexual abuse sent many innocent parents and teachers to prison. The tide began to turn when parents began suing therapists for fabricating such memories, and many of the prison sentences have been overturned by enlightened judges.

The same thing is happening, though on a smaller scale, with fathers falsely accused of sexually molesting a child because of information a child types with a hand held by a facilitator. Lawsuits against FC centers and increasing awareness of the Ouija effect by judges have now almost eliminated such cruel charges. Although a voodoo science seldom completely evaporates, one can hope that the FC farce, involving a mysterious malady more pervasive around the world than Down's syndrome, is finally coming to an end.

REFERENCES

The literature on FC, pro and con, is vast, with hundreds of technical papers. I list here a few popular articles and one book that are not hard to access, and which will reinforce the opinions expressed in this column:

Cowley, Geoffrey. "Understanding Autism." *Newsweek*, July 31, 2000.

Dillon, Kathleen. "Facilitated Communication, Autism, and Ouija." *The Skeptical Inquirer* 17 (Spring 1993): 281–87.

Gorman, Brian J. "Facilitated Communication in America." *Skeptic* 6, no. 3 (1998): 64–71.

Greben, Becky. "Facilitated Communication or Manipulated Miscommunication." *Rocky Mountain Skeptic* 13, no. 6 (1996): 1 ff.

Green, Gina. "Facilitated Communication: Mental Miracle or Sleight of Hand." *The Skeptic* 2, no. 3 (1994): 69–76. Green is director of the New England Center for Autism, in Massachusetts.

Lawrence Erlbaum Associates. *Nonconscious Movements: From Mystical Messages to Facilitated Communication*, 1997.

Mulick, James, John Jacobson, and Frank Kobie. "Anguished Silence and Helping Hands." *The Skeptical Inquirer* 17 (Spring 1993): 270–80.

Osborne, Lawrence. "The Little Professor Syndrome." *New York Times Magazine*, June 18, 2000, 55ff. This article deals with Asperger's syndrome, a milder form of autism involving children with high intelligence who talk like adults, but lack social skills.

Simpson, Richard, and associates. "Effectiveness of Facilitated Communication with Children and Youth with Autism." *Journal of Special Education* 28 (1995): 424–39.

25

Distant Healing and Elisabeth Targ

Elisabeth Targ
Tries mighty hard
To convince everybody that psychics in California can
Heal the sick in Afghanistan.

—a clerihew by ARMAND T. RINGER

I never cease to be amazed by how easily a set of beliefs, no matter how bizarre, will pass from parents to children, and on to grandchildren. I suspect that the vast majority of true believers in every major religion have parents and grandparents of the same faith. It is rare indeed when sons and daughters make a clean break with strongly held fundamental beliefs of their parents.

This was brought home to me recently when E. Patrick Curry, a retired computer engineer, now a consumer health advocate in Pittsburgh, sent me a batch of material about Elisabeth Targ, daughter of the paraphysicist Russell Targ. The team of Targ and his paraphysicist friend Harold Puthoff made a big splash in parapsychological circles in the 1970s. They claimed to have established beyond any doubt that almost everybody is capable of "remote viewing," their term for what used to be called clairvoyance. In addition, they claimed they had validated Uri Geller's psychic ability to remote-view pictures, and his ability to control the fall of dice by PK (psychokinesis). They sat on the fence about Uri's ability to bend spoons and keys because they were never able to capture the actual bending on film. Some parapsychologists called this a "shyness effect."

Russell inherited his psi beliefs from his father, William Targ. When I lived in Chicago I used to visit the father's bookstore on North Clark Street, a store he opened when he was twenty-two. It had a large section devoted to books about the paranormal and the occult. After working for a time as an editor for World Publishing Company, in Cleveland, Targ moved to Putnam in Manhattan where he rose to editor-in-chief. His entertaining autobiography, *Indecent Pleasures*, was published in 1975. At Putnam Targ was responsible for many best-sellers, including Erich von Däniken's notorious *Chariots of the Gods*. (In his autobiography Targ calls it a "quasi-scientific" work on archaeology.) Under his editorship Putnam also published a raft of books about psychic phenomena, such as Susy Smith's *Book of James* in which she reports on channeled messages from the spirit of William James. Targ died in 1999, at age ninety-two. His original name was William Torgownic, taken from his parents when they came from Russia to settle in Chicago where he was born.

William Targ's beliefs in the paranormal trickled down to his son Russell, and now they have descended on Russell's attractive and energetic daughter Elisabeth. Her mother, Joan, by the way, is the sister of chess grandmaster Bobby Fischer. Elisabeth is a practicing psychiatrist with an M.D. from Stanford University, and psychiatric training at UCLA's Neuropsychiatric Institute. Ms. Targ is firmly convinced that persons have the power to use psi energy to heal the sick over long distances even when they don't know the sick but only see their photographs and are given their names.

Elisabeth first participated in psi experiments when she was a teenager. On page ninety-six of *The Mind Race* (1984), a book by Russell Targ and his former psychic friend Keith Harary, Elisabeth is identified as a medical student at Stanford, and an "experienced psi-experimenter and remote viewer." In 1970 she took part in a series of what the authors call successful experiments with a psi-teaching machine. She is said to have recently obtained degrees in biology and Russian.

The authors describe a curious experiment in which Elisabeth correctly predicted in September 1980 that Reagan would win the November election for president. Here is how the test worked.

Ms. Targ's friend Janice Boughton selected four objects to represent the four possible outcomes of the election: Carter wins, Reagan wins, Anderson wins, or none of the above. Each object, its identity unknown to Elisabeth, was put in a small wooden box. Boughton then asked Ms. Targ, "What object will I hand to you at twelve o'clock on election night?"

Elisabeth then predicted the election's outcome by remote-viewing the object she would be given. Her description of the object was white, hollow, conical, with a string attached to the cone's apex. The object that correlated with Reagan's victory was a conical-shaped whistle with a string attached to one end.

Of course six weeks later Ms. Targ had to be handed the box with the whistle. Otherwise, as the book's authors put it, the initial question would have been meaningless.

A similar test of Elisabeth's ability to remote-view a future event involved a horse race at Bay Meadows. On the night before the race, six objects, unknown to Ms. Targ, were assigned numbers that corresponded with numbers on the six horses in the race. As before, Elisabeth was told that at the end of the race she would be given the object that correlated with the winning horse.

Ms. Targ predicted the race's outcome by visualizing something hard and spherical that reminded her of an apple and was transparent. One of the objects was an apple juice bottle. It had been assigned the number on a horse named Shamgo. Shamgo won. Naturally, after the race Elisabeth had to be handed the apple juice bottle to make sense of the experiment.

What a skeptic would like to see would be a transcript of everything Elisabeth said when she was describing the target. Did she say much more than the remarks quoted by her father and his coauthor? If so, there may have been a selection of just those remarks that seemed to describe the target. But I'm only guessing. Also, were there similar tests that failed? One in four, and one in six, are not low probabilities.

There is more about Elisabeth in the book. In May 1982 she and her father conducted a workshop at the Esalen Institute during which successful remote vision tests were carried out with Ms. Targ participating.

Elizabeth Targ is now the acting director of the Complementary Medicine Research Institute (CMRI). It is part of the California Pacific

Medical Center (CPMC). Her institute is devoted to investigating such alternative forms of healing as acupuncture, acupressure, remote healing, therapeutic touch, herbal remedies, meditation, yoga, *chi gong*, guided imagery, and prayer. The institute's literature does not mention homeopathy, reflexology, iridology, urine therapy, magnet therapy, and other extreme forms of alternative healing. Apparently they are too outlandish to merit investigation.

In 1998 Ms. Targ received $15,000 from the Templeton Foundation, an organization established by billionaire John Templeton, an evangelical Presbyterian who showers cash on persons and organizations he thinks are promoting religion. His interest in Ms. Targ's institute springs from her research supporting the healing power of prayer.

In a speech on distant healing that Ms. Targ gave at the Second Annual International Conference on Science and Consciousness, in Albuquerque, New Mexico, April 29–May 3, 2000, she reported that the National Institutes of Health (NIH) now provides funds for research on "distant mental influence on biological organisms." Of more than 135 studies of distant healing on biological organisms, she said, about two-thirds reported significant results. One fascinating study, she added, concerned remote healing of tumors on mice. The study showed that the healers who were *farthest* from the mice had the greatest influence in shrinking the tumors!

Ms. Targ has received $800,000 from the Department of Defense to head a four-year study of the effects of alternative healings on patients with breast cancer. The complementary healings include yoga, guided imagery, movement and art therapy, and others. "We are getting told that we can't study this," she said, "but the beauty of the scientific method is that we can. We can determine if it works—and if so, for whom and how."

CRMI's main achievement so far is a six-month double-blind study of the effects of remote healing on forty patients in the San Francisco Bay area who had advanced AIDS. Forty practicing healers were recruited for the study from healing traditions that included Christians, Jews, Buddhists, Native American shamans, and graduates of "bioenergetic" schools. They were given photographs of the AIDS victims, their first names, and their blood counts.

For an hour every day, over a ten-week period, the healers directed their psi energy to the patients by using prayer or meditation. The experiment was supported by the Institute of Noetic Studies, founded by astronaut Edgar Mitchell, a true believer in all varieties of psychic phenomena, including the powers of Uri Geller, and by New York City's Parapsychology Foundation.

Ms. Targ and three associates reported the results of the experiment in a paper titled "A Randomized Double-Blind Study of the Effects of Distant Healing in a Population with Advanced AIDS." It was published in the prestigious *Western Journal of Medicine* (December 1998). The authors claim that the twenty AIDS patients who received the healing energy (without knowing they had been selected for such treatment), showed significantly better improvement than the twenty patients in the control group who did not receive the energy. As one report summarized the progress of the group receiving the energy, they had "fewer and less severe new illnesses, fewer doctor visits, fewer hospitalizations, and improved mood."

The NIH, through its National Center for Complementary and Alternative Medicine (NCCAM), has provided funding for Ms. Targ to conduct a three-year study of distant healing on 150 HIV patients. The funding for the first year alone is $243,228, with a starting date of July 1, 2000. The NCCAM has also funded a four-year project to study the effect of distant healing on persons with a brain tumor called glioblastoma. The starting date was September 18, 2000, with a first-year grant of $202,596. Both studies, Ms. Targ said, will be double blind. It looks as though Ms. Targ, over the next few years, will be receiving more than two million dollars of government funds for her research on remote healing, the cash coming from our taxes.

Science writer Leon Jaroff, on *Time*'s Web site (January 16, 2002), revealed that Ms. Targ had recently received $611, 516 for one study and $823,346 for another, both awards granted by the National Institutes of Health, a federal organization. Jaroff concludes:

This might all be amusing if Targ's research were not being funded at taxpayer expense by the National Center for Complementary and Alternative Medicine, a controversial branch of the NIH. The least we can demand in a time of growing budget deficits is that NCCAM appoint rational, qualified observers from outside the paranormal and quack communities to monitor the work of some of the eccentrics it so generously endows. Past experience suggests that under such safeguards miracles do not occur.

Ms. Targ is the author of "Evaluating Distant Healing: A Research Review," published in *Alternative Therapies* (Vol. 3, November 1997), and in the same issue, "Research in Distant Healing Intentionality Is Feasible and Deserves a Place in Our Healing Research Agenda." The executive editor of *Alternative Therapies* is Dr. Larry Dossey, who started the distant healing research with his 1993 book *Healing Words: The Power of Prayer and the Practice of Medicine.*

Although Ms. Targ is firmly persuaded that distant healing works, she confesses that no one has any notion of how a healer and healee can be connected over long distances. She closes the second paper just cited with these words: "The connection could be through the agency of God, consciousness, love, electrons, or a combination. The answers to such questions await future research."

Russell Targ's first book, *Mind Reach*, coauthored by Puthoff, is about their tests of remote viewing when they worked for SRI International (then called the Stanford Research Institute). Margaret Mead wrote the book's introduction. Targ's second book, *Mind Race*, was written, as I said earlier, with psychic Keith Harary. His third book, *Miracles of Mind: Exploring Nonlocal Consciousness and Spiritual Healing*, published in 1998 by World Library, is coauthored with Jane Katra, a psychic healer.

The first half of *Miracles of Mind* covers the history of remote viewing, including high praise for Upton Sinclair's book *Mental Radio* about his wife's ability to remote view his drawings. The second half of *Miracles of Mind* is about psychic healing. Targ believes that such healing, especially healing at a distance, is related to the "interconnectedness" of all things by a quantum field such as the nonlocal field of David Bohm's guided wave theory of quantum mechanics.

Miracles of Mind is a strange book. Some chapters are written by Targ, others by Jane Katra. In a few chapters it is hard to tell who is writing. Almost every person engaged in parapsychological research is favorably mentioned, including such far-out paranormalists as Jule Eisenbud, Andrija Puharich, Jeffrey Mishlove, Joe McMoneagle, and many others.

Katra owes an enormous debt to theosophy. She speaks admiringly of Madame Blavatsky, theosophy's founder, as well as England's leading theosophists Annie Besant and Charles Leadbeater. I could hardly believe it, but the book cites (page 94) *Occult Chemistry*, a weird 1898 book by Besant and Leadbeater which describes Leadbeater's clairvoyant probing of the interior of atoms. He is actually credited with having first discovered by clairvoyance that hydrogen has three isotopes!

Miracles of Mind takes seriously such paranormal phenomena as out-of-body travel, near-death experiences, chakras (imaginary energy points in the human body), the Akashik Records (on which all Earthly events are recorded), the visions of Edgar Cayce, and the paranormal powers of Philippine psychic surgeons (to which Katra devotes an entire chapter). There are favorable references to *The Course in Miracles*, a monstrous, vapid tome said to have been dictated by Jesus. Also mentioned without criticism are the powers of Arigo, Brazil's famous psychic surgeon who operated with his "rusty knife" on thousands of patients, following instructions whispered in his left ear by a dead German physician.

Targ credits Jane with having stimulated a seemingly miraculous remission of what had been diagnosed (by whom?) as metastic cancer. "I have been well for the five years since Jane did healing treatments with me," Targ writes. "We will never know if I actually had metastic cancer, or if it was a misdiagnosis. What we do know for sure is that Jane's interactions with me saved me from chemotherapy, which quite likely would have killed me. . . . Did they [his doctors] tell a well man that he had a terminal disease, or did a man with a terminal disease recover through the ministrations of a spiritual healer?" Targ has no doubt that it was Jane Katra who healed him.

The following paragraphs from one of Patrick Curry's letters sum up well the distant healing trend in which Ms. Targ is playing so prominent a role:

> The rise of Elisabeth Targ's distant healing studies is *not* a mere example of defective science leaking into medicine . . . it is a leading wedge of a nascent mystical movement that has been gathering tremendous steam in recent years. The parapsychological enterprise has taken on a new life in its alliance with alternative medicine and

the consciousness movement. What we have is a very productive alliance of parapsychologists, old-fashioned mystics, new-fashioned mystics, and psychedelic mystics that has gotten a major foothold in medicine.

Their presence is extraordinarily strong within NCCAM (National Center for Complementary and Alternative Medicine) and other alternative-oriented sections of NIH (National Institutes of Health). There is a growing presence at dozens of major medical schools, especially Harvard. . . . They have primary devotion not to the ethics of science but to their own belief that they have a mission in serving the New Consciousness. Distortion, and exaggeration of all sorts, are ignored in devotion to their belief in the new paradigm.

As this book goes into page proofs:

On July 18, 2002, Elisabeth Targ died suddenly at the age of forty. Ironically, she died from glioblastoma multiforme, a type of brain tumor she had been researching. She is survived by her father, her husband, and two brothers who are attorneys.

26

The Therapeutic Touch

This essay first appeared in *The Skeptical Inquirer* (November/December 2000).

Millions of ill people around the world, especially in the United States, Canada, Germany, and France, are turning away from mainstream medicine to embrace a bewildering variety of bogus alternative treatments. One of the most bizarre is a growing conviction among American nurses that they possess a mysterious healing power called the therapeutic touch (TT).

The term is misleading because actual touching is not involved. A nurse merely waves his or her hands a few inches above a patient's body. The theory is that everyone is surrounded by an invisible, nonelectric energy field totally unknown to physicists. If a nurse is healthy, energy from her field is transmitted to the ill person. The belief is that this relieves pain, anxiety, and depression, speeds the healing of wounds, and hastens the cure of every imaginable disease.

A vast and growing literature of books and research papers, published mainly in nursing journals, greatly overshadows the few reports of investigations by skeptics. *Therapeutic Touch* is the first book to "debunk" (let's not avoid this word) the wild claims of nurses now so energetically promoting TT.

Therapeutic Touch is a collection of twenty-three papers on TT edited by Béla Scheiber and Carla Selby (Prometheus Books, 2000). Scheiber is a systems engineer, founder, and president of the Rocky Mountain Skeptics, and a member of CSICOP's executive council. Anthropologist Selby is a CSICOP consultant. All but three of the research papers in their valuable anthology provide evidence that the human energy field does not exist, and that the seeming efficacy of TT is fully explained by the placebo effect, by the fact that many ills spontaneously go away, and that the experimental evidence for TT rests on flimsy, poorly controlled testing by true believers who unconsciously bias the results. None of the believers' research has yet appeared in a reputable medical journal.

The book's first paper, by registered nurse Jack Stahlman, is an excellent summary of TT history. Belief that certain individuals have the power to heal by the "laying on of hands" goes back to ancient times. The healings of Jesus involved touching. Kings who ruled by divine decree were believed to have a "royal touch." The Middle Ages and Renaissance swarmed with famous touch healers, and today's fundamentalist and Pentecostal faith healers like to touch the sick. Oral Roberts still firmly believes that energy flows from his hand to cause miraculous healings. The notion of human energy fields is common in Eastern religions, especially in Hinduism and in the Theosophical movement founded by famous charlatan and fake medium Madame Helena Blavatsky.

Stahlman covers the work of three women who have been top drumbeaters for TT: Dora van Gelder Kunz (who died in 1999), Dolores Krieger, and Martha Rogers. Kunz was a theosophist convinced she had clairvoyant powers. She claimed to have read the "auras" (a theosophical term for human energy fields) of thousands of people. She was president of the Theosophical Society for twelve years, chairman of the Theosophical Publishing House, and chief editor of *The American Theosophist*. She is the co-author of a book on chakras (alleged energy centers in the body), and author of *The Personal Aura* (1991).

In the late 1960s Kunz introduced TT to Krieger, a registered nurse now considered the true mother of TT. The former head of the nursing department in New York University's School of Nursing, Krieger taught the nation's first course on TT in 1993 at NYU. She's the author of several books, notably *The Therapeutic Touch: How to Use Your Hands to Help and Heal* (Prentice-Hall, 1979; reprinted by Simon and Schuster, 1992), and *Accepting Your Power to Heal* (Bear and Company, 1993). She has also published many articles in nursing periodicals and is a popular lecturer on TT.

Krieger is a devout Buddhist. Like her friend and mentor Kunz she has been strongly influenced by the writings of British theosophist Charles Leadbeater. Stahlman quotes from one of her books a passage which claims that premature babies who were "declared dead" were revived by TT.

Krieger takes for granted the reality of chakras. She recommends that TT practitioners visualize a color while they are projecting energy. She claims that a skilled touch therapist can store energy in a cotton ball by holding the ball between her hands. The cotton will, she writes, retain the energy for several hours to be used by "healees" to "unruffle" their maladjusted energy field.

TT's most enthusiastic promoter was Martha Rogers, a nurse whose main books are *An Introduction to the Theoretical Basis of Nursing* (1970), and *Science of Unitary Human Beings* (1986). Rogers believed that the human energy field is responsible for ESP, precognition, Kirlian photography, out-of-body experiences, and TT. The Society of Rogerian Scholars issues a journal called *The Rogerian*. [See Jeff Raskin, "Rogerian Nursing Theory: A Humbug in the Halls of Higher Learning," *The Skeptical Inquirer*, September/October 2000. Rogers died in 1994.]

Scheiber and Selby, in several chapters, deal with the explosive growth of TT. There are an estimated 40,000 nurses in the U.S. who now practice the technique with or without the consent of their hospitals. TT is taught in more than eighty colleges.

Like so many other flourishing pseudosciences, TT has not lacked funding. The U.S. Department of Health and Human Services gave $200,000 to a Buffalo nursing center for TT research. The Pentagon's Department of Defense handed $355,225 to the University of Alabama to study the effect of TT on burn patients. NANDA (North American Nursing Diagnosing Association) has approved TT. It is being lauded by the National League for Nursing, the American Nursing Association, and the American Theosophical Society. The Internet has many Web sites defending TT.

The funniest chapter in *Therapeutic Touch* is by philosopher Dale Beyerstein. It records his tireless efforts to locate the elusive "Dr." Daniel Wirth (the "doctor" is for a law degree), who keeps publishing papers on his sensational TT results. Wirth turned out to have a master's degree in parapsychology from John F. Kennedy University, and to be a firm believer in almost every aspect of the psi scene.

Among tests of TT that yielded negative results, two are outstanding. A clever test devised by James Randi for his foundation showed only chance results in efforts by a TT practitioner to detect an energy field with her hands. Randi has a $1.2 million offer, with no takers, for anyone who can detect such a field.

JAMA (*The Journal of the American Medical Association*), in Volume 279, April 1, 1998, published "A Close Look at TT" by Emily Rosa, her mother, and two associates. It reported on a test devised by Emily for a science fair project when she was nine. The test was so simple that it seems incredible that no TT believer ever thought of it. Twenty-one practitioners, convinced they could feel an energy field (it causes a "tingle" in their hands), put their hands, palm up, through holes in an opaque screen. They were asked to tell which hand was under a hand held by another person a few inches above one of the practitioner's palms. The palms were selected by coin flips. Results were at chance level. In a letter to *Time* (May 4, 1998) Krieger predictably scoffed at the test, calling it a "parlor trick," and expressing amazement that *JAMA* would publish such a farce.

Psychologist Ray Hyman, in his wise introduction to *Therapeutic Touch*, likens TT to the phrenology craze, and points out that TT is one of those self-healing belief systems that are unaffected by criticism or negative testing. The burden of proof, he insists, is on the defenders of TT, not on skeptics. He closes with his usual warning that skeptics who attack TT should strive to be fair in their criticisms, and to remember that the ultimate goal is not to debunk but "to promote rational and scientific inquiry no matter what the outcome. . . ."

27

Primal Scream Therapy

This essay first appeared in *The Skeptical Inquirer* (May/June 2001).

Alternative medicines and curious treatments for physical ills are flourishing as never before around the world. The same is true of alternative mental therapies. Every year it seems as if new and outlandish forms of psychiatry appear in books and articles, along with thousands of satisfied patients who provide glowing testimonials about how completely they have been "cured" by the new techniques.

In this chapter I focus on one of the once-popular New Age therapies, the so-called "primal scream" technique discovered and promoted by Dr. Arthur Janov, a California psychologist. Born in Los Angeles in 1924, Janov obtained his doctorate in psychology in 1967 from Claremont College, in Claremont, California. During the Second World War he was a Navy signalman. In 1976 he divorced his first wife, Vivien, who had helped pioneer his work. He later remarried.

Janov and Vivien founded the Primal Institute in Los Angeles in 1970, where some two dozen staffers then practiced primal therapy. Three years later he began publishing *The Journal of Primal Therapy* and the monthly *Primal Institute Newsletter*.

The public first became aware of the new therapy in 1970 when Janov published his first book, *The Primal Scream*. It became an instant best-seller, and the therapy became something of a fad around the world, especially in California. A handsome Janov appeared on the Dick Cavett show. He was interviewed by *Vogue*. John Lennon, Yoko Ono, actor James Earl Jones, and other Hollywood bigwigs praised primal therapy. Sweden aired a long documentary about it.

The basis of primal therapy, which came to Janov like a revelation from on high, is easily capsuled. All neuroses, psychoses, and psychosomatic ills derive from repressed memories of childhood traumas, particularly the violent trauma of being born. This central role of the birth trauma goes back to Otto Rank, a psychotic Vienna psychoanalyst who broke with Freud. Rank traced all neuroses back to a painful birth. He even wrote a book titled *The Trauma of Birth* (English translation 1929), which he dedicated to Freud.

By a series of interrogations—the details of which Janov has kept secret for fear of their being used by untrained therapists—a patient is slowly regressed to childhood. Unconscious memories of incidents which he or she suffered as a very young child start to emerge along with memories of actual birth. When these memories are recovered the ills begin to disappear, though it may take many sessions and much time and money. Moreover, Janov claimed, one's aging process slows down—he once likened his therapy to the Fountain of Youth. Resistance to all diseases increases. In brief, the patient starts to lead a normal, healthy, happy life. Once healed, Janov asserts, a patient will never need therapy again.

The Primal Scream was followed by a spate of popular books with such titles as *The Primal Revolution; The Anatomy of Mental Illness; Primal Man; The New Consciousness* (written with Michael Holden, M.D., then Janov's medical director); *The Feeling Child; The New Primal Scream*; *Prisoners of Pain*; and *Imprints: The Lifelong Effects of Birth Experiences*. All these books are now out of print.

In 1972, when Simon and Schuster published *The Primal Revolution*, it was an alternative selection of several book clubs. A full page ad in *The New York Times Book Review* (November 19, 1972) included a list of ailments primal therapy—and only primal therapy—can cure or alleviate: alcoholism, homosexuality, drug addiction, psychoses, paranoia, depression, and manic-depression. In a similar ad for *The New Primal Scream*, in the same periodical (May 1991), the following ills, all helped or cured by the therapy, are added to the previous list: tension, stress, anxiety, sleep disorders, high blood pressure, cancer, sex difficulties, obsessions, phobias, ulcers, colitis, migraine, asthma, and arthritis.

Not only was Janov convinced that no other form of mental therapy works, but primal therapy must be administered only by workers trained at his institute. Later he speculated that perhaps someday families would learn the technique. This could result in a world with less injustice and no wars. "It would be," Janov is quoted in *Contemporary Authors* (Volume 116), "the only hope if mankind is to survive."

All mental ills, Janov is convinced, result from what he calls "primal pain," a suffering arising from repressed memories of childhood traumas. Illness is a "silent scream." When patients recover their lost memories of early traumas, especially the trauma of birth, they often writhe on the floor, sobbing, and screaming with rage at whatever was done to them or at the violence of their birth.

Such sessions are called "primals." The recovery process is called "primalizing." Primals take place in soundproof rooms with padded floors and walls to prevent patients from injuring themselves while writhing and screaming. The entire process is, of course, faster and cheaper than psychoanalysis, which can go on for years.

Janov was a pioneer practitioner of what later came to be called the "false memory syndrome." During the 1980s and 1990s hundreds of innocent parents and teachers were falsely accused of sexual molestation, frequently of school children. These fake memories were implanted in the patient's mind by well-meaning but self-deceived therapists. Thanks to the valiant efforts of Pamela Freyd, who started the False Memory Syndrome Foundation in 1992, the tide has slowly turned. Judges and attorneys have become aware of how easily such memories can be fabricated, with the happy result that many therapists and quack psychiatrists have lost costly lawsuits, and dozens of innocent adults had their convictions overturned after spending years in prison.

For details about this great psychiatric scandal you can consult the two chapters on it in my *Weird Water and Fuzzy Logic* (1996), or such excellent books as Mark Pendergrast's *Victims of Memory: Sex Abuse*

Accusations and Shattered Lives (1995). The False Memory Foundation is at 1955 Locust Street, Philadelphia, PA 19103–5766.

Janov is not particularly concerned with memories of sexual abuse since any old kind of early childhood trauma will do. Prior to primalizing, patients spend a week in a hotel room without radio, television, or anything to read. They are not allowed to sleep the night before their first session. In his section on primal therapy Pendergrast quotes Janov as saying, "The isolation and sleeplessness are important techniques which often bring patients close to a Primal. Lack of sleep helps crumble defenses."

Of course there is not the slightest reliable evidence that any adult brain harbors repressed memories of birth. Nor, for that matter, any memories of the first one or two years of life, or of prebirth memories of life inside the womb as Janov also believes—a belief he shares with L. Ron Hubbard, Stanislav Grof, and others.[1]

In a letter to *The Skeptical Inquirer* (Fall 1988) Janov canceled his subscription and asked for a refund. He was furious because in the magazine's Winter 1987–88 issue Barry Beyerstein, writing on "The Brain and Consciousness," had called primal scream therapy "suspect."

There have been several tragic spinoffs from primal therapy. In 1971 a Center for Feeling Therapy made its appearance in Los Angeles, founded by two defectors from Janov, Joseph Hart and Richard Corriere. Its techniques included ordering patients to strip and to endure beatings. The center closed in 1980 after losing many lawsuits. Later it was roundly

[1] Grof is a Czechoslovakian-born psychiatrist, 1960s LSD researcher, and paranormalist now living in the United States. SUNY Press has published several of his controversial books. One of Carl Sagan's rare lapses is his unfortunate chapter on Grof in *Broca's Brain*.

In his 1993 book *The Holotropic Mind*, Grof credits LSD with changing him from an atheist into a mystic. He writes (page 18): ". . . we can reach far back in time and witness sequences from the lives of our human and animal ancestors, as well as events that involved people from other historical periods and cultures with whom we have no genetic connection whatsoever. Through our consciousnesses, we can transcend time and space, cross boundaries separating us from various animal species, experience processes in the botanical kingdom and in the inorganic world, and even explore mythological and other realities that we previously did not know existed."

pummeled in such books as *Therapy Gone Mad* by Carol Lynn Mithers (1994) and *Insane Therapy* (1998) by sociologist Marybeth Ayella.

An even uglier spinoff was the rise of "rebirthing therapy," a crazy New Age technique started in the 1970s by one Leonard Orr. The therapy consists of wrapping a patient in blankets to simulate the mother's womb, then pushing pillows onto the patient's face to arouse feelings of labor contractions.

An elderly born-again Christian, Orr lives in his birthplace, Walton, New York, where he runs a rebirthing training center and edits its newsletter. He has written some twenty books. They include *Rebirthing in the New Age* (1977) and *The Healing Power of Birth and Rebirth* (1994). His therapy is closely related to breathing exercises and what he calls the "power of fire." In a trip to India he met a number of yogis who claim to have lived more than two thousand years. One of them, Yogi Babaji, Orr believes to be over nine thousand years old. You can read all about him in Orr's 1992 book *Babaji, The Angel of the Lord.* Somehow Orr manages to combine his Biblical Christianity with India's belief in reincarnation and karma.

In April 2000, in Evergreen, Colorado, a social worker named Connell Watkins and her three associates—none with any training in psychiatry—charged a Durham, North Carolina, pediatric nurse $7,000 for two weeks of therapy on her adopted daughter Candace Newmaker. The girl, ten, was said to be suffering from "attachment disorder," characterized by her inability to form loving relationships. At the culmination of "attachment therapy" the child was wrapped in a flannel blanket, and large pillows were shoved against her face.

Candace cried out repeatedly that she couldn't breathe and was about to vomit, but the therapists kept pushing the pillows and urging her to fight her way out of the "womb" through a twisted part of the blanket. Candace soon stopped crying. A half hour later the therapists unwrapped the blanket. Candace was lying in vomit, not breathing. She died of asphyxiation the next day at a Denver hospital. Watkins and her colleague Julie Ponder were arrested and charged with child abuse resulting in death; their trial began in early April 2000. (For more on this case see "New Age 'Rebirthing' Treatment Kills Girl," *The Skeptical Inquirer* 24[5] September/October 2000.)

If you care to learn more about primal therapy you can read Janov's books, and *A Scream Away from Happiness*, by Daniel Casriel (1972). For attacks on the therapy and its spinoffs see the chapters in Margaret Thaler Singer's *Crazy Therapies* (1996), R. D. Rosen's *Psychobabble* (1977), and Michael Rossman's *New Age Blues* (1979). Rossman's chapter is titled "The I-Scream Man Cometh."

I close on a depressing note. In the spring of 2000 Prometheus Books published Janov's latest work, *The Biology of Love*. In an ad for the book on Janov's Web site, Janov calls it "the most important book of the century." It concerns such questions as, "What makes us humans, the hormones of love, shaping personality in the womb, the nature of feeling, the power of love, the origin of anxiety and depression, the source of addiction and obsessions, sleep and eating disorders, the causes of sexual act out, and many more."

On January 2, 2001, E. Patrick Curry, an articulate consumer health advocate in Pittsburgh, sent Paul Kurtz, founder and head of Prometheus, a strong letter protesting the book's publication. Long an admirer of Prometheus for its willingness to publish books attacking pseudoscience—books other publishers are reluctant to take—Curry urged Kurtz to withdraw the book and issue a mea culpa for the failure of Prometheus editors to recognize Janov's book as bogus psychiatry.[2]

Curry cited an incredible passage on page 319 of *The Biology of Love* that should have been a tipoff to Prometheus editors. Janov reports that a photograph of a primal, in which a patient is experiencing rebirth, shows the fingerprints of the obstetrician miraculously appearing on the patient's legs! "The first time I saw this," Janov writes, "I was as skeptical as I am sure many readers are now. But it happens and is not a chance occurrence."

[2] Paul Kurtz responded to Curry, with a copy to me, on February 7. He noted that despite Prometheus's review process, "We may sometimes err. We are not infallible." But he noted that Prometheus has a long tradition of publishing unpopular books, and criticisms come from virtually every viewpoint. Kurtz said he appreciated Curry's distress with Janov and said he himself was also dubious of "primal scream." He said Prometheus is still committed to a rationalist-scientific agenda but contended that Curry's suggested remedies could be considered suppression. I would consider them damage-control.

If you can believe that, you can believe anything Janov says. To keep up with the doings of what he now calls his Primal Center, in Venice, California, you can check the center's Web site at www.primaltherapy.com. Janov's earlier Primal Institute is now run by his ex-wife, Vivien.

ADDENDUM

The Denver *Rocky Mountain News* (April 5, 2001) published time-lined excerpts from the videotape made during Candace Newmaker's rebirthing session. Read it and weep.

00:00—Therapist Julie Ponder tells Candace to lie down on the navy blue flannel sheet and get into the fetal position.

Ponder: So imagine yourself as a teeny little baby inside your mother's womb and what it felt like. Warm. It felt tight because her stomach was all around you. (Candace is bound in the sheet, the ends twisted above her head and held by Ponder. She is covered by pillows and four adults begin pressing on her.)

01:25—Ponder: What do you think you thought about when you were in there?

I thought I was gonna die.

Ponder: You thought you were gonna die in there?

Yeah.

Jeane Newmaker: I'm so excited. I'm going to have a brand-new baby. I hope it's a girl. I'm going to love her, to hold her and tell her stories . . . I'm going to keep her very safe . . . Every day we'll be together and she'll be with me forever.

(Candace is asked if she believes what her mother is saying.)

Uh huh.

(Candace is asked how that makes her feel.)

Happy.

Watkins: If the baby doesn't decide to be born, she will die. When the baby decides to be born, it's a wonderful thing.

Ponder: So, little baby, are you ready to be reborn?

Uh huh.

Ponder: Come out headfirst. You have to push really hard with your feet. If you stay in there you're going to die and your mommy's going to die.

08:42—Who's sitting on me? I can't do it.

08:53—I can't do it! (Crying). My hands come out first?

Watkins: Sometimes it takes eighteen hours to be born.

09:36—(Screaming) I can't do it. I can't do it! I can't breathe. I can't breathe!

10:16—Whoever is pushing on my head it's not helping. I can't do it. I can't do it. I can't breathe. It's too dark under here. Please quit pushing on my head I can't do it. Somebody's sitting on top of me.

10:50—(Moaning) Somebody's on top of me. Where am I supposed to come out? Right here? Where my finger is?

11:26—I can't do it. (Screams) I'm gonna die.

Ponder: Do you want to be reborn or do you want to stay in there and die?

11:40—Quit pushing on me. Please. (Moaning) Quit squishing my legs. I'm gonna die now. (Screams)

Ponder: Do you want to die?

No, but I'm about to.

12:10—Please, please I can't breathe.

12:30—I can't do it anymore.

12:40—Please quit pushing on me.

13:12—I need some help. Help! Help me please.

Watkins: Are you feeling the contractions, Mom?

Newmaker: I am.

13:43—Where am I to go? Right here? Right here? I'm supposed to go right here? Please. Please. (Screams) OK I'm dying. OK, I'm dying. I'm sorry.

14:31—OK, I'm dying.

14:38—I'm going to die.

15:30—I want to die.

16:08—Can you let me have some oxygen? You mean, like you want me to die for real?

Ponder: Uh huh.

Die right now and go to heaven?

Ponder: Go ahead and die right now. For real. For real.

OK, I'm dead.

Watkins: It's not always easy to live. You have to be really strong to live a life, a human life.

17:07—(Labored breathing) Get off. I'm sick. Get off. Where am I supposed to come out? Where? But how can I get there?

Watkins: Just go ahead and die. It's easier . . . It takes a lot of courage to be born.

18:26—You said you would give me oxygen.

Watkins: You gotta fight for it.

19:50—(Candace vomits) OK, I'm throwing up. I just threw up. (Vomiting) I gotta poop. I gotta poop.

21:24—Uh, I'm going in my pants.

Ponder: Go ahead.

Watkins: Stay in there with the poop and vomit.

23:22—Help! I can't breathe. I can't breathe. It's hot. I can't breathe.

Newmaker: I'm so excited to have this baby. I'm waiting for you, to love you and hold you . . .

Ponder: Scream, Candace.

No.

Newmaker: Baby, I love you already. I'll hold you and love you and keep you safe forever . . . Don't give up on your life before you have it . . .

32:25–33:44—Jack McDaniel repositions himself on a pillow over Candace's head.

Ponder: Candace? (No response) (Takes another pillow from Newmaker.) She needs more pressure over here so she can't . . . so she really needs to fight.

Watkins: Getting pretty tight in here.

Ponder: Yep . . . less and less air all the time.

35:39–40:00—Ponder and McDaniel reposition themselves again.

Ponder: She gets to be stuck in her own puke and poop.

Watkins: Uh huh. It's her own life. Quitter.

40:01: No. (This is Candace's last word.)

McDaniel: Mama got you this far, now it's up to you.

Watkins: Candace is used to making her life everybody else's problem. She's not used to living her own life.

Ponder: Quitter, quitter, quitter, quitter, quit, quit, quit, quit. She's a quitter.

(Watkins leaves, Newmaker leaves. McDaniel takes Watkins's place. Watkins returns.)

McDaniel: This baby doesn't want to live. She's a quitter.

(Watkins tells McDaniel and St. Clair to take a break.)

(Ponder and Watkins discuss someone who is stressed, then chitchat about their dream homes and a million-dollar property nearby that is being remodeled.)

Watkins: Let's talk to the twerp.

(They unwrap Candace.)

01:09:53—Watkins: Oh, there she is sleeping in her vomit.

28

Eyeless Vision

This essay first appeared in *The Encyclopedia of the Paranormal*, edited by Gordon Stein (Prometheus, 1996).

As all magicians know, but most people do not, it is almost impossible to blindfold someone so securely that he or she cannot peek down the side of the nose. "For whatever care may be taken to deprive a person of sight in this way," wrote Robert Houdin, the great French conjuror in his *Memoirs* (1858), "the projection of the nose always leaves a vacuum sufficient to see clearly." Harry Houdini, who took his name from Houdin, put it this way in *A Magician Among the Spirits* (1924):

> Putting cotton on the eyes and covering it with a handkerchief is now used by amateurs. There is not the slightest difficulty in seeing beneath such a bandage, sometimes over it, and the range of vision can easily be determined by a test. In Paris I saw a magician, named Benoval, who had his eyes glued together with adhesive paper, on top of it cotton was placed, and over the cotton a handkerchief, but he danced around bottles and burning candles without any difficulty.

Ignorance of blindfold deception methods has been especially widespread in the investigations of persons who claim they can read or tell colors without using their eyes. At times such "eyeless vision" is said to be clairvoyance. More often it is attributed to a mysterious ability, unknown to science, to "see" with fingertips, toes, the forehead, or some other part of the body.

Spiritualist Conan Doyle, writing about the ability of Andrew Jackson Davis, an early American seer, to see while his eyes were blindfolded, explained his power this way in *The History of Spiritualism* (New York: Doran, 1926, Vol. 1, pp. 43–44):

> At first, the gift was used as a sort of amusement in reading the letters or the watches of the assembled rustics when his eyes were bandaged. In such cases all parts of the body can assume the function of sight, and the reason probably is that the etheric or spiritual body, which possesses the same organs as the physical, is wholly or partially disengaged, and that it registers the impression. Since it might assume any posture, or might turn completely round, one would naturally get vision from any angle, and an explanation is furnished of such cases as the author met in the north of England, where Tom Tyrrell, the famous medium, used to walk round a room, admiring the pictures, with the back of his head turned towards the walls on which they were hung. Whether in such cases the etheric eyes see the picture, or whether they see the etheric duplicate of the picture, is one of the many problems which we leave to our descendants.

The classic crank work on what is now called "dermo-optical perception" (DOP) was *Vision Extra-Rétinienne* by the prolific French novelist, poet, and dramatist Jules Romains (1885–1972). It was first published in Paris in 1919. Charles Kay Ogden, an English scholar who collaborated with I. A. Richards on the creation of Basic English, and who co-wrote with Richards an influential book on semantics, *The Meaning of Meaning*, translated Romains's book into English. Putnam's published it here in 1924 with the title *Eyeless Vision*.

Romains, supremely ignorant of how to prevent cheating, investigated a raft of French women who claimed they could read while "securely blindfolded." He became convinced that human skin possesses organs sensitive to light, and that these organs are most sensitive at the tips of fingers. He believed this because his blindfolded subjects ran fingers over print while reading it. They could not read in total darkness and performed poorly in dim light. Romains thought the mucous lining of the nose was sensitive to colors because when his blindfolded subjects identified colors held in front of their face, they had a tendency to tilt back their head and sniff.

Tilting back the head, of course, makes a nose peek easier. By today's standards, Romains's controls were incredibly amateurish. Adhesive tape would usually be crossed over closed eyes, then cotton would be pushed into spaces alongside the nose. As any French magician could have told him, it is easy to shove cotton aside with concealed thumbs during the act of adjusting a blindfold. A ridiculously simple way to prevent such peeks is to put an aluminum box over a subject's head, resting it on padded shoulders and fitting snugly around the neck, with air holes at top for breathing. Romains thought of bibs under chins, but the idea of a box over the head never occurred to him.

Romains's book stimulated American magicians to devise a variety of eyeless vision acts. In Chicago Harlan Tarbell used adhesive tape to seal down his eyelids before covering them with a conventional blindfold. His technique for gaining sight is explained in the *Tarbell Course in Magic*, Vol. 6 (Louis Tannen, 1954). Tarbell credits Romains with inspiring his act, and refers to a similar performance in the early 1920s by a woman using the stage name of Shireen. Houdini (1924) says that Shireen, "securely blindfolded," would hit target bull's-eyes with a rifle. Other conjurors invented other techniques, such as taping silver dollars or powder puffs over each eye before a blindfold was added.

Dozens of ingenious blindfolds have been devised for the magic trade. They are first placed over a spectator's eyes to prove their opaqueness, then a subtle adjustment of the cloth gives the performer clear vision. A simple dodge with an unprepared blindfold is to pleat it several times, then allow spectators to convince themselves they cannot see through it. Later the same cloth is pleated a different way. Opposite corners are each folded toward the central diagonal. The final form looks the same as before, but now the performer has a clear view ahead through the single thickness of cloth between the pleats.

Nose peeks with sleep blindfolds are easily obtained by wrinkling the forehead to raise the cloth slightly. Other types of blindfolds, such as goggles with opaque lenses, allow peeks over the bridge of the nose. In recent years magic shops have carried metal blindfolds that fit too snugly around the nose to allow peeks, yet there is an undetectable way to gain straight-ahead vision.

Trick blindfolds are used by all magicians and psychics who perform what the trade calls a "blindfold drive." With eyes "securely blindfolded," the magician drives a car through busy streets. The Israel "psychic" Uri Geller, in younger years before he stopped doing conventional magic, performed blindfolded drives and also featured eyeless vision in his stage magic acts.

The most famous performer of eyeless vision was Kuda Bux, a Moslem magician from Kashmir, India. His act is described in Harry Price's *Confessions of a Ghost Hunter* (London: Putnam & Co., 1936). Assistants from the audience would first cover Bux's eyes with large globs of dough, then yards of cloth would be wound turbanlike around his entire head from forehead to chin. In the act of adjusting the cloth where it passed over his nose, Bux's thumbs, hidden behind fingers, would push alongside his nose to form channels that ran to the inside corners of his eyes. The dough would soon harden, leaving permanent slots through which he could read print and writing on blackboards. Bux also performed blindfolded drives in cars and on bicycles.

Life (April 19, 1937) published three pages about a thirteen-year-old Glendale, California, boy named Pat Marquis. "Securely blindfolded," Pat is shown shooting pool, playing Ping-Pong, and fencing. He could name playing cards with his eyes covered, and drive a car. In all the photographs Pat is seen tilting his head back to obtain the usual nose peek. Before performing, the lad would go into a trance and assume the personality of his previous incarnation, an eleventh-century Persian.

Pat's talents were discovered and promoted by a British brain surgeon, Cecil Reynolds. Dr. Reynolds claimed that the boy saw through a spot in his forehead. When Joseph B. Rhine, Duke University's famed parapsychologist, tested Pat later in 1937 the boy wore goggles with lenses made opaque. Rhine caught Pat cheating by peeking over the bridge of his nose.

In 1962 all of Russia was transfixed by reports that Rosa Kuleshova, a tiny twenty-two-year-old epileptic patient at First City Hospital in Lower Tagil, could read print while "securely blindfolded." Sensational articles about her appeared in Russian newspapers and popular magazines. *Time* (January 25, 1963) picked up the news and featured a photograph of Isaac Goldberg, a Russian psychiatrist, observing a blindfolded Rosa while she ran a finger over a newspaper page. Rosa's head is tilted back in a sniff position.

Naive, gullible scientists were completely taken in by Rosa's crude deceptions. Papers giving glowing reports about her amazing talent appeared in Russian scientific journals. Soon scores of other DOP claimants popped up all over the Soviet empire. "The fingers have a retina!" exclaimed biophysicist Mikhail Smirnov. Rosa began showing how she could read print covered by cellophane or glass. Scientists were dumbfounded when a green book was suddenly flooded with red light and a blindfolded Rosa shouted, "The book has changed color!" She was able to name cards and describe pictures by sitting on them.

USSR, an English-written magazine later called *Soviet Life*, devoted four pages to Russian DOP readers (February 1964). This inspired *Life* to send a reporter to Russia and to hire science writer Albert Rosenfeld to tell the Russian story in nine pages. Titled "Seeing Colors with the Fingers," the report included a page of symbols, printed in different colors, so *Life* readers could give themselves DOP tests. Rosenfeld bought it all. So did Sheila Ostrander and Lynn Schroeder in their breathless chapter on Russian DOP, in their best-seller *Psychic Discoveries Behind the Iron Curtain* (Englewood Cliffs, N.J.: Prentice-Hall, 1970).

The enthusiasm of Russian-born Gregory Razran, then the respected head of the psychology department at Queens College, New York, was boundless. Here are some excerpts from what he told Rosenfeld:

> In all my years, I can't remember when anything has had me more excited than this prospect of opening up new doors of human perception. I can hardly sleep at night. And I can hardly wait to get

started on my own experiments. When the word gets around, and when people begin to realize what a major scientific breakthrough has been achieved, we're going to see an explosive outburst of research in this field. The results are bound to be revolutionary. To see without the eyes—imagine what that can mean to a blind man!

For all we know, this may turn out to be some entirely new kind of force or radiation that has gone undetected—and unsuspected—until now. There is nothing ridiculous about this idea. After all, the history of discovery in this field is that we sense something first, then we go out into the sea of energy that surrounds us and look for whatever it was caused the sensation. Since we have not experienced this particular sensation before, we didn't know there was anything to look for. Now that we have started to look, there is no telling what we may find.

We know that many lower organisms have light sensors all over their bodies. If man has some kind of light-sensing mechanism on his skin which he has been unaware of—a sensing mechanism perhaps left over as a vestige of some earlier stage of evolution—it should not come as a great surprise.

In 1964 Razran thought he had discovered an American counterpart to Rosa. She was Linda Anderson, a fifteen-year-old girl from Billerica, Mississippi. Her father, Arthur Anderson, claimed to be a hypnotist. After many mesmeric sessions, he said, his daughter had developed the power to read while "securely blindfolded."

After appearing on several TV shows, Linda received nationwide publicity in November 1964, when she offered to aid the police of Lowell, Massachusetts, in finding Kenneth Mason, a five-year-old boy who had disappeared. Linda told police the boy was not in the Merrimac River, where he was thought to have drowned, but was "in a house" outside the state.

A test of Linda's eyeless vision was arranged at the laboratory of Biometrics Research, part of New York's State Department of Mental Hygiene, in Manhattan. A report of what happened, by the department's Joseph Zubin, appeared in *Science* (Vol. 147 [February 26, 1965]: 985). Razran headed the team of observers. He was furious when he learned that Zubin had invited magician James Randi to be present.

The tests were a disaster. Although a blindfold was taped around its edges, Linda constantly managed to create tiny apertures by tensing facial muscles. Under Randi's supervision, zinc ointment was used to plug each aperture when it appeared. This always prevented Linda from "seeing" until she managed to contort her face enough to create a new opening. As Zubin writes, "The white ointment made it much easier for the experimenters to detect the occurrence of a chink."

At one point, when Linda was unable to loosen the tape enough to get a peek, the experimenters left the room so she could confer privately with her father. Unknown to them, Randi had left a tape recorder running. Played later, they heard Linda say, "Daddy, Daddy. I can't see! What shall I do?"

"Just a minute, dear," the father answered. After a moment of silence, Linda said, "That's better, Daddy, I can see now."

When the scientists returned they saw a chink through which Linda was reading nicely.

Randi told me in a letter that Linda had a strongly concave nose that allowed her to peek with either eye over the bridge of her nose as well as through chinks on other sides of the blindfold. She would hold an open newspaper, arms outspread, and pretend to be looking at the center of a page, when actually she was reading a column far to one side. Pinhole openings, Randi reminded me, give high-resolution images.

Linda constantly chewed gum, and the resulting strong facial contortions would loosen the tape. She rubbed her cheek, adjusted her hair, and complained about the pressure of the tapes. Her eyes, under the blindfold, she insisted, were always closed. Razran kept saying to Randi, "But how could an adolescent girl learn to do magic tricks that professionals have to practice for years?" When he finally became convinced the girl was cheating, he vowed to resign from the committee investigating her.

The exposure of Linda's cheating did not prevent her father from continuing to have her tested, hoping for a validation by a scientist that would help make his daughter famous. James A. Coleman, a physics professor at American International College, in Springfield, Massachusetts, arranged a test on February 12, 1965, offering a hundred-dollar prize to Linda if she passed his challenges. The judges were three businessmen from the area, two of them amateur magicians who knew all about nose peeks.

The outcome was as disastrous as before. Coleman wrote an amusing factual account of the occasion that ran as a four-part article in *Yellow Jacket* (February 18; March 12, 19, and 26), his college's student newspaper. Only the names were altered. Linda is called Wispy Espy, her father Arty Andy, and himself Professor Boldman. Linda is described as petite, cute, pretty, and seemingly deeply religious. She claimed that her talent came from God, and that when she grew up she wanted to be a minister and do missionary work.

> Little Wispy Espy radiated innocence and holiness as she told the news people, angelically, "It was one day last July, and I was in church, looking up at the cross. When I looked down at my hymn book, I saw light shining right through it and I actually could see through the book."
>
> The news people listened, enraptured. They were bewitched by Espy's sweet little voice, her cherubic little face and her sanctimonious air.

Linda's behavior followed the patterns described by Zubin. Her makeup made it difficult for tape to remain in place. By muscular contortions, eye twitching, and gum chewing she constantly produced tiny holes around the blindfold. Whenever a chink appeared, Coleman would plug it with ointment. "It was very difficult," he writes. "It was like trying to plug up holes in a rowboat while someone is peppering it with a machine gun." Whenever a hole was plugged, Linda was unable to see.

The newspeople present, as well as the judges, were convinced that both Linda and her father were charlatans. Coleman then surprised everyone by introducing Randi (he calls him Dandi), who regaled the reporters with a graphic description of similar events in Zubin's laboratory.

A month later the body of Kenneth Mason was found in the Merrimac River, not out of state in a house as Linda had predicted. In September a man confessed to having murdered the boy by drowning. I have no idea what subsequently happened to Linda or her father.

In 1973 Razran drowned at age seventy-two while swimming off the beach at St. Petersburg, Florida, where he lived. His *New York Times* obituary (September 2) cited his major work *Mind in Evolution: An East-*

West Synthesis (Boston: Houghton Mifflin, 1971), but made no mention of his interest in DOP. I do not know if he ever doubted his premature acceptance of DOP, or whether Rosenfeld ever changed his mind. I had lunch with Rosenfeld in 1963 when he was a science editor of the *Saturday Review*. . . . The *New York Times* (October 11, 1970) reported that Rosa was tested by five scientists who caught her peeking under her blindfold. She died in 1978 after touring for several years with a circus where she gave DOP performances. Among the many new DOP claimants was Lena Bliznova, age nine, who staggered Russian parapsychologists by her ability to read print, "securely blindfolded," with fingers held several inches off a page. She separated black chess pieces from white and also read print with her toes. Reading with the toes touching pages on the floor allows for easy nose peeks without tilting.

Ninel (Lenin backwards) Kulagina began her meteoric rise as Russia's most famous psychic by performing eyeless vision. When she was tested by a group of competent investigators, she totally failed. They first allowed conditions that permitted peeks, then tightened controls to prevent it. Later she was caught using invisible thread and concealed magnets to perform feats of psychokinesis. (See *Scientific American*, March 1965, pp. 57–58.)

John Davy in "Can You See With Your Fingers?" in London's *Observer* (February 2, 1964), quoted Romains as complaining bitterly that Russian and American scientists, completely ignoring his findings, had merely "repeated one twentieth of the discoveries I made and reported."

In America Richard P. Youtz, chairman of the psychology department at Barnard College, became the most heralded American defender of DOP. He did not accept the ability to read print with fingers, but became convinced that fingertips could distinguish colors. His belief was based on tests made with . . . a housewife in Flint, Michigan, who seemed able to distinguish colors on test cards and pieces of cloth while "securely blindfolded."

Newsweek covered Youtz's research favorably (December 30, 1963). The *New York Times* praised his investigations in two articles by Robert K. Plumb: "Woman Who Tells Color by Touch Mystifies Psychologist" (January 8, 1964), and "Sixth Sense Is Hinted in Ability to 'See' with

Fingers" (January 26). The *Times* even ran an unsigned editorial, "Can Fingers 'See'?" (February 6, 1964).

The *New York Times Magazine* (March 15, 1964) published "We Have More Than Five Senses," by Leonard Wallace Robinson. Youtz's work, he wrote, arouses "the very strong possibility that some human beings may be able to see with their fingertips." For my critical letter about this article, and Youtz's long reply, see the *New York Times Magazine* (April 5 and 26, 1994).

Unfortunately, after Youtz responded to criticism by tightening his controls, the woman's ability evaporated . . . "Housewife Is Unable to Repeat Color 'Reading' With Fingers" was the headline of a *Times* story February 2, 1964. Youtz blamed the failure on cold weather that affected his subject's fingers. He tested her once more in warm weather, again with negative results. This time he blamed the failure on fatigue.

I was never able to persuade Youtz to put a box over the housewife's head, although he did devise a plywood box with armholes through which she inserted her hands. A grant from the Institute of Mental Health, which had partly funded his work with his subject, allowed Youtz to continue DOP research with 135 women students at Barnard. He reported that 20 showed rudimentary ability to sense colors by DOP.

Psychologist and magician Ray Hyman, of the University of Oregon, was an observer at one of Youtz's 1964 tests of the housewife. In a letter to me he described Youtz's procedures as "unbelievably sloppy." His color cards had tape on them that varied in thickness from card to card. Some cards had smooth edges, others had serrated edges. During the trials Youtz and the housewife carried on a constant interchange of small talk in which Youtz, quite unconsciously, provided cues about how she was doing. In reports Youtz called his tests "double blind," but when Hyman pointed out that Youtz was always aware of each color card and how he placed it in his light-proof box, he replied that he was using the term in a weaker sense. He meant that after he put several test cards in the box and shut the door, his subject then mixed the cards so neither of them knew which was which!

One of Youtz's tests involved three phosphorescent crosses. He would expose one cross to light, then place it in the box along with two others to see if his subject could pick out the glowing one. Hyman observed that in every trial Youtz put the illuminated cross in first, and had he been in the chair he could have located it by sound alone. However, she scored chance level on this experiment. Either she was not alert enough to take advantage of the sound cues, or she was reluctant to perform knowing she was being observed by a person knowledgeable about deception.

A strong attack on Youtz's tests of Barnard students was made by psychologist Robert Buckhout, of Washington University, St. Louis, in his paper "The Blind Finger," *Perceptual and Motor Skills* (Vol. 20 [February 1965]: 191–94). In addition to finding flaws in Youtz's "Summary of Aphotic Digital Color Sensing: A Progress Report," presented at an April 18, 1964, meeting of the Eastern Psychological Association, Buckhout tried to replicate Youtz's experiment using forty undergraduates. The test material consisted of squares of construction paper, in ten different colors, mounted on black paper and covered with plastic film. The experiment was genuinely double-blind. Good scorers received a dollar for each follow-up session. Overall results were at chance level, and there was no significant difference between male and female scoring.

As late as 1968 Youtz was still defending DOP in "Can the Fingers See Colors?" in *Psychology Today* (February 1968). The housewife had by then faded from the scene as quickly as Linda. Youtz died in 1986, never doubting the reality of DOP.

A twenty-one-year-old woman identified only as "S" was tested in 1965 for DOP by J. Zachary Jacobson, B. J. Frost, and W. L. King, all then at Dalhousie University, in Quebec, Canada. They reported their findings in "A Case of Dermooptical Perception," *Perceptual and Motor Skills* (Vol. 22 [1966]: 515–20). The woman wore tight-fitting welder's goggles, their lenses covered by adhesive tape. Loosely balled Kleenex was placed over each eye under the goggle. Her arms were then inserted through holes in a plywood box painted black within, but open on the side opposite the holes so test cards could be put inside. The experiment took place in a photographic darkroom. The cards were rectangles cut from red, yellow, green, and blue construction paper, half of them covered with plastic, and half covered with glass.

After a card was presented to S, she would feel it and call out the color. The card would be removed, the set of cards shuffled, then another card placed in the box. When a light within the box was off, S scored at chance levels, but when the light was on she scored significantly above chance.

In a letter to me Ray Hyman pointed out two major flaws in the procedure: (1) randomizing of cards by hand shuffling rather than by an efficient randomizing method; and (2) after each guess S was told whether it was correct or not. This feedback allowed for a tactical method of cheating. For example, S could make a slight nick on the edge of a card with her fingernail and so recognize the card when it came to her again.

Even without such marking, there may have been slight differences in the cards that could be felt. We are not told, for example, how the plastic and glass were attached. The paper says the glass was "clamped" to each card. Does this mean some type of metal clamp was used, or was the glass "clamped" by tape around the edges? In either case, subtle differences in the clamps or tape could have enabled a clever subject to recognize a card of known color by carefully feeling it.

This loophole could easily have been eliminated by not telling S how she did on each guess. The experimenters might argue that if S cheated, why did she fail when the lamp in the box was off? The answer could be that because she was aware of the experimenters' conjecture that colors radiate different degrees of heat when illuminated, she went along with this theory. . . . S was not informed when the light was on or off, but surely she could tell from the lamp's heat. In any case, I know of no replication of this experiment.

The notion that blind people have the ability to "see" colors by touch is probably as old as civilization. James Boswell, in his life of Samuel Johnson, reported that on Easter Sunday (April 19, 1772),

> General Paoli and I paid him a visit before dinner. We talked of the notion that blind persons can distinguish colours by the touch. Johnson said, that Professor Sanderson mentions his having attempted to do it, but that he found he was aiming at an impossibility; that to be sure a difference in the surface makes the difference of colours; but that difference is so fine, that it is not sensible to the touch. The General mentioned jugglers and fraudulent gamesters, who could know cards by the touch. Dr. Johnson said, "the cards

used by such persons must be less polished than ours commonly are."
Nicholas Sanderson (or Saunderson) was a blind mathematician at Cambridge who wrote a two-volume *Elements of Algebra.*

James Hogg, writing on "Colour Blindness," the *Popular Science Review* (London, Vol. 2 [1863]: 407–509), said: "I know and have met with very many instances in the totally blind able to distinguish every variety of colours by the delicacy of their sense of touch: they tell me there is a sensible difference in the degree of heat conveyed to the point of the finger."

Numerous papers have been written here and abroad about teaching the blind to distinguish colors by DOP, some even written by blind persons who claim such an ability. It is good to remember that "blind" is a fuzzy word. Many people who are legally blind, and can honestly claim they are totally unable to see, nevertheless have retinas capable of distinguishing darkness from light and seeing colors dimly. To determine whether a blind person is actually sensing colors by DOP, controls to prevent seeing must be tight. It is not enough for such persons to make tests on their own, with eyes uncovered, because they may be sensing colors on an unconscious subliminal level.

The possibility that differences in heat reflected by colors might be detectable by fingers was suggested by the research of Walter L. Makous, of IBM's Watson Research Center, in Yorktown Heights, New York. He reported his work in "Cutaneous Color Sensitivity: Explanation and Demonstration," first issued as an IBM Research Paper (October 29, 1965), and reprinted in *Psychological Review* 73 (1966): 280–84). To prevent his subjects from seeing the test material, he used an opaque screen under which arms could be thrust, and an opaque apron tied about the neck and thrown over the head. "If there is reason to suspect a subject of trickery," he writes, "an aluminum box of the type Gardner recommends can be used." Whether he actually used such a box is not clear.

Makous found that fingers could distinguish between a polished metal plate that reflected light, and a plate painted black to absorb light. He speculated that there might be sufficient heat differences between red and blue objects to enable a blind person, with practice, to discriminate between them.

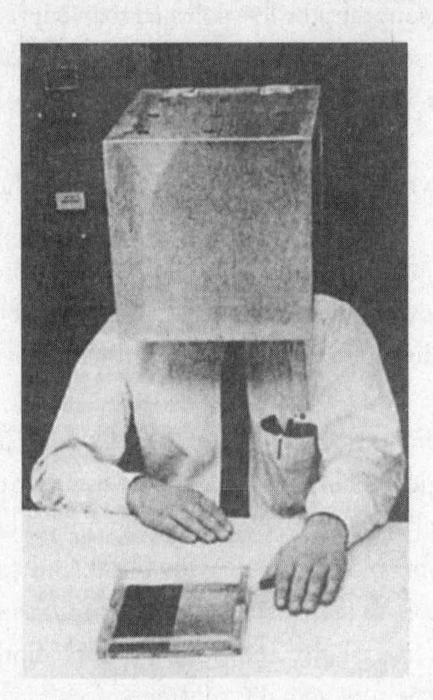

Psychologist Walter L. Makous, wearing a ventilated aluminum box designed by amateur magician Martin Gardner to prevent "nose peek," sits before a stripped-down version of a test plate whose radiant emissivity in a light-tight box can vary as much as 0.5 degrees Celsius on the subject's hand as it moves from polished to black surfaces.

An experiment similar to Makous's was conducted by Carroll Blue Nash, a zoologist and parapsychologist then at St. Joseph's University, Philadelphia, where he had founded a parapsychology laboratory. Forty-three subjects, wearing boxes over their heads, were tested for their ability to discriminate by touch between sheets of red and black construction paper. All but one subject scored above chance on uncovered sheets. Thirty scored above chance on sheets taped to a plastic covering. Hits were at chance level on sheets covered with glass. Nash believes that a difference in the intensity of infrared radiation from the papers accounted for the positive results. One defect of his procedure was telling a subject after each guess whether it was correct or not.

Nash reported this experiment in "Cutaneous Perception of Color with a Head Box," in the *Journal of the American Society for Psychical*

Research (Vol. 65 [January 1971]: 83–87). Now retired, Nash wrote more than 120 papers for psi journals, many co-authored by his wife, Catherine Stiffler Nash. He is the author of two books on what he calls "psiology."

Even with a head box precautions must be taken. Unless the box fits closely around the neck, downward peeks are possible. Straight-ahead vision can be obtained by a small convex mirror attached to a belt buckle or palmed. Perhaps the simplest way to exclude all peeks is to tape an oval piece of opaque cloth securely over each eye.

There also is the possibility of secret signaling by an accomplice posing as an observer. Psychic mountebanks have been known to refuse demonstrations unless a relative or friend is present to send them good vibrations. The "vibrations" can be a secret signaling by a variety of subtle methods. For example, he or she can have a switch in the tip of the shoe for sending beeps to a tiny receiver in the psychic's rectum, a place not likely to be inspected.

Although many magicians and psychics continue to perform eyeless vision, no one since Kuda Bux has featured such acts over a longer period than Ronald Coyne, a Pentecostal evangelist living in Tulsa. For forty years he has been wowing fundamentalist congregations around the nation and overseas.

At age seven Coyne lost an eye after it was severely injured by barbed wire. With his good eye "securely blindfolded" by a folded handkerchief taped around the sides Coyne uses his plastic eye to read print on anything handed him. He even removes his artificial eye and "reads" with the empty socket. The demonstration is then followed by faith healing and rousing altar calls.

Coyne began his long career as a youngster. Now middle-aged, he is still astounding believers with his "miracle vision." From his Tulsa headquarters you can obtain *When God Smiled on Ronald Coyne* (Tulsa, Okla.: Ronald Coyne Revivals, 1958), a thin paperbound book written by his mother, and a long-playing record of one of his early revival services.

In 1979 an outbreak of eyeless vision in China was kicked off by a report in the *Sichuan Daily* (March 11, 1979) about a twelve-year-old boy who could read documents with his ears. Soon dozens of children all over China were making equally outrageous claims. "Securely blindfolded," they read with their fingers, nose, scalp, abdomen, and even their buttocks. They read messages on paper crumpled and stuck in their ears or under their armpits. One little girl had only to touch a document with the tip of her pigtail.

China's respected *Nature Journal* began trumpeting these amazing talents, calling them instances of EBF (Extraordinary Body Functions). Newspapers around the world picked up the mania. (See the *Los Angeles Times*, March 2, 1980.) Parapsychology journals hailed the discovery of China's talented children. Parapsychologist Robert A. McConnell, in chapter 2 of his *Parapsychology and Self-Deception in Science* (privately printed, 1983), included a paper by Dr. C. K. Jen, a Chinese-born physicist with a Harvard doctorate, that extolled the Chinese ear and armpit readers as genuine.

In 1981 sociologist Marcello Truzzi and his friend, parapsychologist Stanley Krippner, flew to China to investigate. Truzzi reported their findings in "China's Psychic Savants," *Omni* (January 1985). They not only found no evidence of DOP, but they also caught the children flagrantly peeking. In 1982 the Chinese Academy of Science investigated the children and reached the same conclusion. Such criticism has had no effect on Chinese parapsychologists. To this day they continue to make wild claims for psychic children who bamboozle them with simple magic tricks.

I should add that one of the secrets of eyeless vision acts is to get a peek at something when it is handed to you and held momentarily below your nose where no head tilt is necessary. You quickly memorize the words so that when the writing, say on a business card, is held directly in front of your face, where a nose peek is impossible, it can then be "read." Psychic researchers are constantly taken in by this dodge, claiming later that nose peeks were ruled out because the print was held on eye level while being "read."

The earliest Russian paper on DOP known to me is "A Rare Form of Hyperesthesia of Higher Sensory Organs," by psychiatrist A. N. Khovrin, in *Contributions to Neuropsychic Medicine* (Moscow, 1898). Khovrin describes the ability of a Russian woman he calls Sophia to read print with her fingers and to taste with her forearms.

Reverend I. Platts, in his 952-page *Encyclopedia of Natural and Artificial Wonders and Curiosities* (New York: World Publishing House,

1875), devotes six pages to a teenage girl called Miss M'Avoy who lived on the east side of Liverpool. An attack of hydrocephalus was said to have caused total blindness. Yet when she was "securely blindfolded" by a shawl placed over her eyes, she could read fine print by passing a finger over it. Edith Sitwell, in *English Eccentrics*, calls her Margaret McAvon. She was born in Liverpool in 1800 and became blind in 1816. Her talent created a sensation in Liverpool.

Platts quotes from a long article about Margaret in the Liverpool *Mercury* that describes a reporter's observations in 1816: "One circumstance, which has created doubt and suspicion," the reporter writes, "is that if any substance, for instance a book or shawl, be interposed between her eyes and the objects she is investigating, she is much embarrassed, and frequently entirely baffled."

Miss M'Avoy also demonstrated finger reading while wearing goggles, their lenses covered with opaque pasteboard. She also could name the colors and tell the shapes of objects placed against the back of her hand. A reporter for the Liverpool *Advertiser* is quoted as saying that on one occasion Miss M'Avoy read print when a visitor stood behind her and pressed both hands over her eyes.

In the *British and Foreign Medical Review* (April 1845) a Dr. Noble reports on a boy he calls Jack whose eyes were "bound down by surgeons with strips of adhesive plaster, over which were folds of leather again kept in place by other plasters." Jack could read anything handed him. A Manchester surgeon, however, after his own eyes were covered exactly the same way, found that by working his facial muscles he could produce a chink above the bridge of his nose. It was then suggested that Jack demonstrate his ability after his eyes were sealed shut with shoemaker's wax. He strongly objected, and sure enough, after the wax was applied he could not see. William B. Carpenter describes this case in his book *Mesmerism, Spiritualism, Etc., Historically and Scientifically Considered* (New York: Appleton, 1877). "Jack now plainly saw," Carpenter writes, "even with his eyes shut, that his little game was up."

True believers are seldom convinced by negative evidence. Alfred Russel Wallace, a spiritualist who, like Conan Doyle, accepted all forms of psychic phenomena, castigated Carpenter in his article "Dr. Carpenter on Spiritualism," in the *Popular Science Monthly Supplement* (November

1877: 435–50). Carpenter's ignorance is blamed for his stubborn refusal to recognize genuine clairvoyance. Wallace cites the case of Thomas Laycock, a boy in Plymouth who in 1846 read fluently after his eyes were covered with three layers of adhesive plaster. This positive proof of clairvoyance, Wallace says, was recorded in the *Zoast* (Vol. 9: 84–85). The *Zoast* was the monthly organ of British mesmerism and paranormal research. It ran through thirteen volumes before expiring in 1856.

Psychic researcher Richard Hodgson, in Britain's *Journal of the Society for Psychical Research* (June 1994) tells of testing a boy in England called "Dick the Pit Lad." He seemed to read while "securely blindfolded," but Hodgson found that when he bandaged his own eyes the same way he could peek easily.

A teenage Frenchman named Alexis Didier created a stir in France and England with his DOP performances in the mid-1840s. Frank Podmore, in *Modern Spiritualism* (London: Methuen, 1902, Vol. 1, p. 144) reports that Didier would first go into a trance, then leather pads were placed over each eye. A handkerchief was then tied diagonally over each eye, a third handkerchief tied horizontally across both eyes, and cotton wool pushed into the interstices.

> But a writer in the *Morning Chronicle* tells us that he had himself been bandaged by a friend in the same way, and managed to read distinctly. It was noticed, moreover, by several persons that Alexis contorted his face both during and after the process of bandaging; that he frequently touched or fidgeted with the bandages; that he held the objects to be looked at at curious angles, and changed their position, as if trying to get a better view.

Podmore cites numerous references in British periodicals to Didier's performances in England.

Dozens of articles defending DOP appeared in occult periodicals during the 1960s and 1970s. See, for example, *Fate* (July 1963, May 1964, May and July 1976, and September and October 1967); *Psychic* (June 1971); and *New Realities* (Vol. 1, no. 4 [1977]). A translation of a French work, *The Paranormal Perception of Color*, by France's top DOP researcher, Yvonne

Duplesis, was published in 1975 by the Parapsychological Foundation of New York.

For more on DOP, see my paper "DOP—A Peek Down the Nose," in *Science* (Vol. 151 [February 11, 1966]), reprinted with additions in my *Science: Good, Bad and Bogus* (1981). Youtz's letter objecting to this paper ran in *Science* (May 22, 1966: 1108). See also his letter in *Scientific American* (June 1965: 8–9). My column in the *Skeptical Inquirer* (Summer 1994) is devoted to Ronald Coyne.

I suggest that from now on DOP stand for "Dermo Optical Peeking."

References

Coleman, James. "The ESP Girl." *Yellow Jacket*, February 18, March 12, 19, 26, 1965.

Coyne, Mrs. R. R. *When God Smiled on Ronald Coyne*. Tulsa, Okla.: Coyne Revivals, 1952.

Duplesis, Yvonne. *The Paranormal Perception of Color*. New York: Parapsychological Foundation, 1975.

Gardner, Martin. "DOP—A Peek Down the Nose," in *Science: Good, Bad and Bogus*. Amherst, N.Y.: Prometheus Books, 1981.

Ostrander, Sheila, and Lynn Schroeder. *Psychic Discoveries Behind the Iron Curtain*. Englewood Cliffs, N. J.: Prentice-Hall, 1970.

Romains, Jules. *Eyeless Vision*. Translated from the French by I.A. Richards. New York: Putnam, 1924.

Rosenfeld, Albert. "Seeing Colors with the Fingers." *Life*, June 12, 1964, 102ff.

29

Magic and Psi

This essay first appeared in *The Encyclopedia of the Paranormal*, edited by Gordon Stein (Prometheus, 1996).

Conjuring is the art of performing what appear to be miracles for the purpose of entertainment. Because magicians are the world's experts on the art of deception, it is one of the scandals of psychic research that investigators, except on rare occasions, will not seek the aid of knowledgeable conjurors when they test psychics who perform feats unexplainable by natural laws.

There are endless examples of this reluctance. Philosopher-psychologist William James made no effort to consult magicians in his investigations of Mrs. Piper and other mediums. Although Conan Doyle was for a short time friendly with Houdini, he never asked Houdini or any other magician to attend a séance with him. Indeed, Doyle came to believe that Houdini himself was a powerful medium. How else explain his miraculous escapes?

Scores of high-standing parapsychologists, who believed that the Israeli magician Uri Geller could bend metal with his mind, had no impulse to ask magicians to monitor his feats of ESP and PK (psychokinesis). I once urged writer Charles Panati to take a magician with him when he had his first encounter with Geller. He did not do so. Geller blew Panati's mind with elementary trickery, and Panati edited an embarrassing anthology called *The Geller Papers*. Psychiatrist Jule Eisenbud never wanted a knowledgeable magician to watch Ted Serios, the Chicago bellhop, project his thoughts onto Polaroid film before Eisenbud wrote an entire book about Serios's curious paranormal ability.

One obvious secret shared by psychics and magicians is the ability to tell convincing lies. Magicians repeatedly lie, either directly or indirectly, but audiences, knowing they are being flimflammed, do not mind. A magician will say, "I'll place your selected card under your foot" when the card has already been switched for another card. Psychic charlatans tell similar fibs, but psi researchers, predisposed to believe a psychic is honest, seldom question what the psychic says. If a test is not going well and a psychic attributes this to a headache, researchers will not doubt what he or she says, and the failed experiment will never be recorded. When a medium claims she has no memory of anything said during a trance, true believers never question it.

Hundreds of instances of how psi performers lie could be given. For example, a psychic asks for a house or car key, and as soon as he takes it he will give it a secret bend. Most keys, especially if long and deeply notched, are not hard to bend with strong fingers. The key is then placed in the owner's fist to be held high over his head while he repeats, "Bend, bend, bend!" Does he not feel the key getting warmer? Of course he does, because he is squeezing his fist.

Now comes the subtle fabrication. The psychic asks to see the key. "No, nothing has happened," he lies. "Let's try once more."

The psychic walks to the other side of the room. He commands the key to bend. When the viewer lowers his fist and opens it, lo and behold, the key is twisted. By now he has forgotten that the psychic momentarily held the key. He will swear to his dying day that the key bent while *in his fist*, and the psychic was twenty feet away. At no time, he will tell others, did the psychic touch the key.

This scenario illustrates another important secret that magicians and psychics share—the inability of laymen to remember what they see. They simply do not know what to look for, and there are subtle ways performers plant false memories in their minds. Every magician has had the amusing experience of hearing someone later describe one of his tricks in such a way that it could not possibly have been done.

One of the simplest dodges of great slate-writing mediums in the glory days of spiritualism was to "accidentally" drop an examined blank slate on the floor, then reach down and pick up a duplicate slate with chalk writing already on its underside. There are documented occasions when trained investigators completely forgot a slate had been dropped, testifying later that they carefully examined a blank slate and it was never out of their sight while messages materialized on it.

A psychic who claims he can predict in advance what someone will draw may do this as follows. He will pretend to sketch his prediction on a pad but actually make only scratching sounds with a fingernail. The blank pad is placed facedown on the table. The viewer will then draw on another pad a picture of, say, a house. The psychic now pretends he did not do well. He will pick up his pad, apologize for the crudeness of his sketch, then replace the pad facedown. While momentarily holding it he has quickly sketched a house, using what magicians call a thumb writer. After more conversation the psychic asks his guest to pick up the pad. On it is a picture, crudely drawn, yet unmistakably a house!

A few days later all memory that the psychic picked up his pad before the picture was revealed will have been totally erased from the observer's mind. He will later insist that the psychic "never touched the pad" until he, the observer, turned it over. Because he heard the psychic's pencil as he drew the picture, the fact that the psychic glanced at his drawing seems so entirely irrelevant that it slips forever from the observer's memory.

Suppose a psychic asks you to go into another room and draw something on a sheet, fold the paper, and seal it in an opaque envelope he gives you. You bring him the envelope. He starts to put it in his jacket pocket, then suddenly changes his mind and withdraws it. "The envelope really should be left here on the table," he says, "where you can keep your eyes on it." He has already secretly switched envelopes. Days later you will have totally forgotten that for an instant the envelope was out of sight. You may not even remember that later the psychic excused himself to visit the bathroom. All that sticks in your memory is that the sealed envelope was placed on the table in full view at all times while the psychic accurately duplicated your drawing. Momentarily putting the envelope in a pocket seems too irrelevant to be remembered. Did he hand the envelope back to you after his success? Well, no. Come to think of it, he tore it up and tossed it in a wastebasket. After all, why would you want to open it since you knew what you had drawn?

Today's parapsychologists do not take seriously paranormal events performed in the dark, but in the past this was a universal cover for mediums. Why spirits were afraid of light was never clear. What would you think of a magician who said, "I have a wonderful trick to show you. I'm going to conjure a flower out of thin air, but first we have to make the room so dark you can't see anything." Lights go out, then back on, and he is holding a daffodil in his hand.

Magician John Mulholland, whose *Beware Familiar Spirits* is an excellent exposé of the tricks of mediums, liked to point out that when a magician materializes an elephant on a well-lit stage, before a thousand viewers, a believer in psi will be totally unimpressed. It's all done, he will shrug, with trapdoors and mirrors. That same person will sit for an hour in a pitch-dark room with a medium. When the lights go on, if there is a rose petal on the table, he will be overwhelmed with awe.

John Beloff, England's best known parapsychologist, firmly believes that the famous medium D. D. Home could actually levitate and float near the ceiling the room was totally dark. Home would leave a mark on the ceiling provided to prove he was really up there. I once asked Beloff in an article if he truly believed that if someone had suddenly turned on the gaslight Home would have dropped to the floor. To my amazement, Beloff said yes. For some unknown reason, he wrote, darkness is essential to certain paranormal phenomena!

Why is it that mediums no longer perform miraculous feats in darkness? The answer is easy. In days before electricity it took some time for a skeptic to get up, walk across a room, and turn on a gaslight. Today an unbeliever can produce a flashlight and the jig is up.

Magicians who perform what in the trade is called "mentalism" have invented hundreds of ways by which information can be transmitted secretly from one person to another. Yet when two psychics appear on the scene to demonstrate their telepathic abilities, it almost never occurs to a psi researcher to ask a knowledgeable magician to watch. Today's miniaturization technology has produced easy forms of wireless communication. A secret accomplice can transmit beeps by a switch in a shoe. What psi researcher will insist that a psychic, claiming to receive telepathic messages from a sender in another room, undress and be examined by a doctor who looks for tiny devices in a tooth or up a rectum?

Even assuming such a careful search is made, would the investigator be smart enough to examine the psychic's clothing, then watch her carefully after examining body and dress to make sure she does not pick up a receiver from someone or someplace? Houdini would have his body thoroughly searched by a doctor before he went behind a screen to escape from a handcuff. All he needed to do was take the proper pick from a compartment in the screen. At other times he would go to the bathroom to find the picks in a toilet's flush box or under the rim of a commode.

One subtle communication system is known to magicians as a time-delay code. Let's say a sender in one room concentrates on a digit from 1 through 9 when it is flashed on a screen by a random number generator. In another room the receiver tries to guess the digit by ESP. The sender presses a button when he starts to concentrate on the number, and presses it again when he is finished. Sender and receiver wear watches with second hands. The time interval between start and finish signals conveys each digit. If a psi researcher is unfamiliar with such a code, would he ever suspect it?

It is a common myth that magicians accomplish many of their tricks by rapid movements. "The hand is quicker than the eye." The opposite is true. Fast motions of a hand call attention to the hand. If a conjuror reaches high in the air to produce a cigarette, the other hand may be doing something essential. This is called misdirection. (In gambling circles it is called "shade.") If for example a card hustler wants to exchange a fair deck for a marked one, known in the trade as cold-deck switching, he may "accidentally" knock over a glass of beer. In the confusion of mopping up the beer, the switch is made. One famous American performer featured an illusion that required an assistant off stage to remove something from a large box placed near the wings. This occurred exactly at the moment a scantily clad young woman entered the stage on the opposite side.

Psychics who perform magic tricks are skilled in misdirection. To give one example, while all eyes are on a hand waved above a compass needle, a leg is lifted under the table with a magnet strapped above the knee, or a foot touches the table's underside with a magnet in its shoe. Another spot for a secret magnet is under the tip of a shirt collar. While the psychic clenches his fist above the compass, he keeps lowering his head, as if to get a better view, until the magnet in his collar causes the needle to gyrate. Many a naive researcher has been convinced that a psychic moved a compass needle by PK because when he examined the psychic's hands he found them empty!

One enormous advantage psi tricksters have over magicians is the unshakable belief of psi researchers. In most cases psychics, who refuse to be tested by skeptics, give as their reason that skepticism inhibits their powers. Even with believers they may be unable to work their wonders unless conditions are exactly right. A psychic may take thirty minutes to bend a spoon, waiting for a chance to prepare the spoon properly or have an accomplice do it. If conditions are never right for a trick, he can always say he is not feeling well. On the other hand, a magician must succeed every time he performs. It would kill his career if he ever said to an audience, "I'm sorry, but there are skeptics among you and the vibes are not right. There will be no magic this evening."

Who would book a magician if he performed only when he felt like it, or if he took half an hour to produce a rabbit from a hat? When Merv Griffin introduced Geller on his TV show, he made the mistake of assuming that because Geller's tricks had failed on occasion, it proved that he was not in fact a magician because magicians' tricks always work. Although traps for catching psychics with smoking guns are easy to devise, psi researchers almost never stoop to such deceptions. Do they not indicate distrust that is sure to influence a psychic's powers? Yet such traps are often the only way to prove that fraud is being used. In 1983 Harry Collins, a British sociologist and psi believer, was impressed by the ability of British children to bend cutlery by PK. Yet no one had actually seen the metal while it was bending. This was dubbed the "shyness effect." In an effort to catch the twisting on videotape, Collins put the children in a room where they thought they were not observed. Actually, they were secretly taped through a one-way mirror. When Collins saw the tape he was astounded. The children were seen doing all the bending by physical force. One-way mirrors and secret videotaping are excellent ways to trap psychics. When Walter Levy, in J. B. Rhine's laboratory, was caught cheating on an experiment with laboratory animals, it was done by suspicious colleagues watching through a wall peephole.

Ted Serios accomplished his "thoughtography" by palming into a paper tube, which he held in front of the camera lens, a tiny optical device. It was explained in detail by two magicians in *Popular Photography* (October 1967). Had Jule Eisenbud known anything about magic he could easily have trapped Ted by grabbing his wrist immediately after a picture was snapped, and examining the paper tube before Ted had a chance to "go south" with the device. Such a thought never crossed Eisenbud's trusting mind.

Nina Kulagina, Geller's Russian counterpart, used invisible thread to move matches across a table and to float Ping-Pong balls. It is possible that the thread was manipulated by her husband in a side room. Any magician present could have recognized the method at once and simply passed a hand through the space where the thread went before it could be drawn out of the room.

Felicia Parise thoroughly bamboozled parapsychologist Charles Honorton by using invisible thread stretched between her hands when she pushed a pill bottle across her kitchen counter. Had Honorton known anything about thread magic (books about it are sold in magic supply houses, along with strong thread so fine that it cannot be seen in bright daylight), he would have examined Felicia's hands while the bottle was gliding. He did no such thing. Instead he later examined the top of the counter! When on the phone I told him Felicia had used invisible thread, Honorton said this was impossible because the bottle had moved away from her! In his total ignorance of conjuring, he could conceive only of a thread attached to the bottle that Felicia would drag toward her!

If you suspect that a member of your club is cheating in poker games, whom would you call in to catch him in the act? A physicist? A lawyer? A policeman? A parapsychologist? Obviously you would hire someone knowledgeable about ways to cheat at cards. If a psychic does tricks that look like magic, the last person to validate his work is a scientist.

As any magician will tell you, scientists are the easiest of all people to fool. It is because they are used to investigating a universe that always plays fair. "God may be subtle," Einstein liked to say, "but he is not malicious." In a letter he put it this way: "Nature hides her secrets through her intrinsic grandeur but not through deception." As I have said before, electrons and laboratory animals don't cheat; humans do.

Scientists reason logically, but magic operates along illogical paths, by techniques totally unlike the laws of nature. If a magician puts a ring on a string held at both ends, a scientist wants to examine the ring and the string for secret openings because he knows that such linkages violate laws of topology. But there are no secret openings. The string is ordinary and the ring may be borrowed. Scores of ways to perform this trick have been invented, not one of which a scientist could suspect. The methods of magic are *sui generis*, irrational, bizarre, completely outside the scope of a scientist's experience. This explains how such intelligent men as Oliver Lodge, William Crookes, Alfred Russel Wallace, and many other scientists of the recent past became ardent Spiritualists. It explains how so many scientists and engineers today can be deceived by psychic mountebanks.

What is the moral to all this? It is simply that when miraculous psi phenomena are tested in a laboratory, no investigation is worth a dime unless a knowledgeable magician is there to observe. Not just any magician, because some conjurors, though famous stage performers, know less about magic than well-informed amateurs. It must be a magician thoroughly familiar with what is known as close-up magic—magic performed not on the stage but within a few feet of spectators.

It is not enough for a knowledgeable magician to serve as consultant before or after an experiment. He has to be there. No psi investigator, unless thoroughly trained in the strange, semi-secret art of conjuring, knows what to look for or how to design a trap that will expose a clever charlatan. If he thinks he is capable of detecting fraud without the aid of someone who knows close-up magic, he is guilty of both an inflated ego and the sin of willful ignorance.

30

How Mrs. Piper Bamboozled William James

This essay first appeared in *The Encyclopedia of the Paranormal,* edited by Gordon Stein (Prometheus, 1996).

Mrs. Leonora Piper (1859–1950) was the most famous direct-voice medium in American history. Unlike other mediums of the time, she never produced physical phenomena. She merely went into trances during which spirits of the dead took over her vocal cords or seized her hand to write what they dictated.

Leonora Evelina Simonds was born in Nashua, New Hampshire. She never finished high school. When twenty-two she married William J. Piper, a Boston bookstore clerk and ardent Spiritualist. Their first home was in Boston's Beacon Hill section where they raised two daughters, Minerva and Alta. Mr. Piper died in 1904. Alta wrote *The Life and Work of Mrs. Piper* in 1929. Both daughters became professional musicians.

Mrs. Piper was tall, stout, and handsome, with blue eyes and brown hair, good-natured, self-possessed, matronly, modest, and shrewd. At age eight, Alta writes, her mother's head was injured by a mysterious blow over her right ear, followed by a voice that said, "Aunt Sara not dead, but with you still." Later Leonora learned her aunt had died.

In her mid-twenties Leonora fell into a trance during a visit to J. E.

Mrs. Leonora Piper

Cocke, a blind medium. After awakening she was told that a young Indian girl named Chlorine had spoken through her. Soon Mrs. Piper was giving her own private séances.

William James's mother-in-law was so impressed by a Piper séance that she recommended the medium to William. He began attending her sessions and sending relatives and friends. Alice, James's wife, became convinced Mrs. Piper was channeling discarnates. As one of James's biographers put it, Alice was "credulous" where William was mostly "curious."

After her fame spread abroad, Mrs. Piper made three trips to England under the auspices of the Society for Psychical Research (SPR). Her most eminent supporters there were Spiritualists Frederic Myers and physicist Sir Oliver Lodge. In America her most enthusiastic defender was James Hervey Hyslop, professor of logic and ethics at Columbia University.

When Mrs. Piper entered a trance her pupils dilated, she moaned and sobbed, her eyes rolled upward, and her ears wiggled violently. Later this transition became calmer. Her breathing would slow, and she usually snored throughout her deep sleep. As her trance ended she would often weep and mutter expressions of pain, pleasure, or disgust. During one trance she ignored a small cut James made on her wrist. She was undisturbed when a needle was forced into her hand and when a French investigator stuck a feather up her nose.

Mrs. Piper's voice always altered when different controls took over. Men spoke like men, children like children. Foreigners had accents characteristic of their backgrounds.

When Mrs. Piper moved to Arlington Heights, near Boston, her séances were held upstairs in what she called her Red Room because its wallpaper and furnishings were red. A clock was kept illuminated after the room was darkened. Was this so she could know when to end a séance?

After Chlorine, Mrs. Piper's controls were Martin Luther, Commodore Cornelius Vanderbilt, Longfellow, George Washington, Lincoln, J. Sebastian Bach, English actress Sara Siddons, and Loretta Pachini, a young Italian. After they ceased coming, the next and most famous control was an eighteenth-century physician Dr. Phinuit (pronounced Phinuee), who had died of leprosy. Oddly, his name sounded very much like the French physician Albert G. Finett (pronounced Finee) who had been the control of the blind medium who launched Mrs. Piper on her career.

Phinuit said he came from Metz. Efforts to find evidence in Metz of a doctor who once had lived there were fruitless, even though Phinuit gave his birth and death dates. He spoke English with a stage-French accent, but could not speak French. He claimed he had lived so long in an English-speaking colony that he had forgotten French except for a few common phrases.

Mrs. Piper's next major control was known as G. P., the initials of George Pelham. The name was a pseudonym to conceal the identity of George Pellew, a writer who died in 1882 after falling off a horse in New York City. A month later he turned up as one of Mrs. Piper's controls.

Pellew was a friend of Richard Hodgson, a British psi researcher who came to Boston in 1887 to serve as secretary of the American Society for Psychical Research (ASPR), and to edit its journal. (He died in 1909 of heart failure while playing handball.) When Pellew began coming through Mrs. Piper, Hodgson was so skeptical that he hired detectives to shadow Mrs. Piper and her husband for several weeks to make sure they were not researching information about his friend. Finding no such evidence he became convinced he was indeed conversing with his deceased friend.

In later years Mrs. Piper's voices were replaced by automatic writing. While in trance her right hand rapidly scribbled messages. Frequently she pressed so hard the pencil broke. For a while she spoke and wrote simultaneously. On several occasions three discarnates came through, one speaking, one writing with one hand, one writing with the other. (Mrs. Piper was strongly ambidextrous.) Her right hand also served as a strange telephone. If sitters wanted to ask the control a question, they held the hand close to their mouth and spoke with a loud voice.

In 1896 Mrs. Piper's spirit contacts became a group called the Imperators—critics called them the Imposters—who had earlier been controls of William Stainton Moses, a famous British medium. Immortals on a higher plane, they chattered constantly about God, heaven, and angels. In 1905 they were supplanted by the dead Richard Hodgson.

What persuaded so many clients that spirits indeed spoke through Mrs. Piper? It was the astonishing amount of information they provided. Even William James believed she got this data paranormally, though he was never convinced it came from the dead or was picked up telepathically from sitters. He preferred to think she was tapping some vast, transcendent Mother Sea.

Records of Mrs. Piper's séances show plainly that her controls did an enormous amount of what was called "fishing," and today is called "cold reading." Vague statements would be followed by more precise information based on how sitters reacted. Mrs. Piper usually held a client's hand throughout a sitting, sometimes holding the hand against her forehead. This made it easy to detect muscular responses even when a sitter was silent. Moreover, her eyes were often only half closed, allowing her to observe reactions.

Dr. Phinuit had a habit of babbling what James called "tiresome twaddle" while he shamelessly fished. Caught in an error, he would profess deafness and leave. His ignorance of science and literature was monumental, yet he was well-informed about hats and clothing.

Books about Mrs. Piper by believers seldom mention her information failures. She once told James a ring had been stolen, but it was found in James's home. Three times Phinuit tried vainly to guess the contents of a sealed envelope in James's possession, even though the doctor supposedly contacted the dead woman who wrote the letter. Frederic Myers also left a sealed letter for mediums to read, but Myers himself, speaking through Mrs. Piper, could not read it. Myers's wife, in a letter to the *London Morning Post* (October 24, 1908) said she and her son found nothing of value in any of the messages from her husband that came through Mrs. Piper.

Although cunning cold reading may account for most of Mrs. Piper's hits, I believe she had other tricks up her sleeves. She constantly saw friends and relatives of clients. A vast amount of personal information can emerge in the give and take of séance conversation, to be fed back to sitters in later séances. Because sitters believed Mrs. Piper when she claimed she recalled nothing during a sitting, it never occurred to them she could be lying. She also may have obtained information from conversations of clients in a sitting room awaiting the start of a session. Visitors often chatter away, revealing facts easily overheard by a medium or her relatives on the other side of a wall.

Obtaining facts about prominent persons is not difficult. Obituaries can be checked. Courthouses contain birth and marriage records, real estate sales, and so on. Reference books abound in biographical data that sitters often swear a medium could not possibly know.

In her hagiography, Alta speaks of servants, nurses, and governesses in their large home. In 1885 they had an old Irish servant whose sister worked for a prominent Beacon Hill family often visited by James's mother-in-law. Yet when the mother-in-law sat with Mrs. Piper, William was flabbergasted to learn that her controls had named members of his family!

Mediums in a city know one another. Those who patronize one medium usually visit others. At the time there were scores of mediums in Boston, forming a network of scoundrels who passed information freely back and forth.

The strongest indictment against James is that not once did he even think of devising a simple "sting." Even his invalid sister, Alice, a skeptic of all things psychic, initiated a trap. When James asked for a lock of her hair to give to Mrs. Piper—"vibrations" from personal items helped her make contact with appropriate spirits—Alice slyly sent him a lock from a deceased friend. Here is how Alice described the hoax in a letter to William in 1886:

> I hope you won't be "offended" . . . when I tell you that I played you a base trick about the hair. It was a lock, not of my hair, but that of a friend of Miss Ward's who died four years ago. I thought it a much better test of whether the medium were simply a mind-reader or not, if she is something more I should greatly dislike to have the secrets of my organisation laid bare to a wondering public. I hope you will forgive my frivolous treatment of so serious a science.

Did James report back to Alice what Mrs. Piper's controls said about the hair? If so, his letter has not survived.

Alice's opinion of James's spiritualist friends could not have been lower. In one letter to her brother she called Myers an "idiot." Here are her candid thoughts as she jotted them in her diary a week before she died of breast cancer:

> I do pray to Heaven that the dreadful Mrs. Piper won't be let loose upon my defenceless soul. I suppose the thing "medium" has done more to degrade spiritual conception than the grossest forms of materialism or idolatry: was there ever anything transmitted but the pettiest, meanest, coarsest facts and details: anything rising above the squalid intestines of human affairs? And oh, the curious spongy minds that sop it all up and lose all sense of taste and humour!

In his essay "The Will to Believe" James likens anomalies in science to the white crow that falsifies the assertion "All crows are black." "My own white crow is Mrs. Piper." In the midst of all the humbug is "the presence . . . of really supernormal knowledge . . . in strong mediums this knowledge seems to be in abundance, though it is usually spotty, capricious and unconnected."

"If spirits are involved," James wrote in a later essay, they are "passive beings, stray bits of whose memory she [Mrs. Piper] is able to seize." James was aware of how Mrs. Piper's controls shamelessly fished for data, yet he could not avoid thinking her messages were "accreted round some originally genuine nucleus." Belief in psi phenomena has lasted so long through the centuries, he argued foolishly, that there must be something to it. Here is what I consider the most stupid remark in all of James's writings:

> When a man's pursuit gradually makes his face shine and grow handsome, you may be sure it is a worthy one. Both Hodgson and Myers kept growing ever handsomer and stronger-looking.

In 1901 Mrs. Piper wrote an extraordinary two-and-a-half-page article for the *New York Herald* (Sunday, October 20). Its headline: "I Am No Telephone to the Spirit World." She was, she said, retiring as a medium. She wished to be "liberated" from the ASPR, for which she had served as an "automaton" for fourteen years. She desired freedom for more congenial pursuits. She wanted to tell the world that nothing she had said or written while in trance could not be explained as coming from her unconscious or obtained by telepathy from sitters or persons elsewhere. "I must truthfully say that I do not believe that spirits of the dead have spoken through me. When I have been in a trance state . . . it may be that they have, but I do not affirm it."

Yet Mrs. Piper did not retire. For two more decades she conducted séances. New controls included George Eliot, Julius Caesar, and Madame Guyon, a seventeenth-century French mystic. When Conan Doyle visited her in 1922, she had lost all her powers.

Alta attributes her mother's temporary loss of power from 1911 to 1914 to the harsh experiments conducted by psychologist Granville Stanley Hall, then president of Clark University, and his assistant Amy B. Tanner. So strict were their conditions that Imperator, one of Mrs. Piper's controls, issued an ultimatum that her "power must be withdrawn for a time in order to repair the machine." (Mrs. Piper's body was known as "the machine.") A detailed account of this investigation is given in Tanner's *Studies in Spiritism* (1910), reprinted in 1994 by Prometheus Books.

Tanner and Hall approached Mrs. Piper with open minds. Tanner not only believed in ESP, but also did not rule out the possibility of spirit communication. Hall was more skeptical, though as a youth he had accepted

Spiritualist claims, and as an adult continued to believe in immortality. In 1909 he and Tanner conducted six sessions with Mrs. Piper, all recorded verbatim in Tanner's book. They ended their research persuaded that Mrs. Piper was perhaps not a charlatan, but a classic case of a person with multiple personalities who emerged from the unconscious during trances. Had the book accused her of fakery, Hall and Tanner would have been open to libel suits. If you read carefully between the lines, there are suggestions that in many ways Mrs. Piper practiced deceptions.

One indication of this was that, unlike other cases of persons with subliminal personalities, Mrs. Piper's trances never occurred spontaneously. They never began when she was alone or asleep. Yet whenever a sitter paid for a séance, she had no difficulty going into a trance. Moreover, persons suffering genuine trance seizures do not go in and out of them in theatrical ways calculated to impress audiences.

Another suggestion of conscious acting was the incredible role played by Mrs. Piper's right hand. The fact that the hand functioned as a telephone to spirits was, as Hall remarks, a miracle in itself. Sitters were asked to hold the hand, its palm close to their mouth, and to speak with a loud voice as if on a long distance phone call. Occasionally the hand would explore a sitter's face or body.

Did the hand require shouting because Mrs. Piper was getting deaf? On the contrary her hearing was extremely acute. As Hall reports, she reacted to everything audible—"noise on the street, the rustle of clothing, the sitter's position, and every noise or motion."

By insisting that sitters address the hand in a loud voice, a strong impression was created that Mrs. Piper was "as much out of the game as if she were dead." If the hand could not hear voices in low tones, surely Mrs. Piper could not hear the conversation of sitters. Convinced that the sleeping Mrs. Piper could hear nothing, sitters felt free to talk to one another. Later they would not even remember what they had said. When information from such whispered conversations came out in later séances, or even in the same séance, they would be amazed.

Hall explained it this way:

> Here then is a wide and copious margin in which suggestion can work. Never in our own or in other Piper sittings was any full record

> kept of what her interlocutors said. Still less have involuntary exclamations, inflections, stresses, etc., been noted, and even the full and exact form of questions is rarely, if ever, kept, while the presence of a stenographer which we proposed was objected to. Thus, unlimited suggestions are unconsciously ever being given off to be caught and given back or reacted to in surprising ways. If this method be a conscious invention on her part it shows great cleverness and originality, and if it be a method unconsciously drifted into, its great effectiveness could in fact be scientifically evaluated only by prolonged experiments in which a normal person should simulate her very peculiar kind of sleep. In fact, it often seemed that only her eyes were out of the game, and all her mental and emotional powers were very wide awake. A little practice convinced me that it is not hard to feign all this, and yet I am by no means convinced that she acted her sleep-dream, although that this could be done with a success quite equal to her own I have no shadow of a doubt. If this is the case she is, of course, fraudulent, but if some of her faculties are really sleeping it is a unique and interesting case of somno-scripticism as her former practice of speaking instead of writing was of somno-verbalism, for both are species of the same genus of somnambulism. That Mrs. Piper-Hodgson's soul is awake and normal, our last sitting gave abundant evidence when she seemed to quite fall out of the Hodgson role and became angry.

Hall's greatest scam was presenting Hodgson, Mrs. Piper's main control at the time, with fictitious information and names. Hall had met Hodgson only once, but he pretended they had been old friends. Hodgson reciprocated by calling Hall "old chap" and by remembering a wealth of events and discussions that never took place. Hall invented a Bessie Beals. Hodgson had no trouble locating her on the "other side."

In a hilarious final séance, Hall openly revealed to Hodgson his many deceptions, and did his best to convince him he was not really Hodgson but only a fragment of Mrs. Piper's mind. Hall and Tanner could not hold back laughter at Hodgson's discomfort and evasions. "To oblige me," Hall said, repeat the words 'I am not Hodgson.' " Hodgson refused. "No, I am Hodgson." Hall ordered him to "fade away." Hodgson said he would go only when ready. After he finally left, one of the Imperators came on with "May the blessing of God rest on you."

Although Mrs. Piper always insisted she never recalled anything that transpired during a séance, Hall noticed a growing coldness in her attitude toward him and his assistant. After this final disastrous scene, she betrayed no hint of knowing how funny it had been.

Mrs. Piper died in 1950, at age ninety-one, and almost totally deaf. Until then she had occupied an old apartment in a Boston suburb with her youngest daughter, Minerva. The last article about her while she lived was "America's Most Famous Medium," by Murray Teigh Bloom in the *American Mercury* (May 1950). Here is how he closed the article:

> Few in the comfortable, old-fashioned apartment house know that the very old lady who occasionally goes out for a stroll with her nurse or gray-haired daughter, is the simple Yankee housewife whose work once convinced leading scientists of two countries that there was indeed life after death.

References

For a fuller account of Mrs. Piper's career, see my two-part article in *Free Inquiry* (Spring and Summer 1992) and the references cited at the end.

Bloom, Murray Teigh. "America's Most Famous Medium." *American Mercury* (May 1950): 578–86.

Brandon, Ruth. *The Spiritualists*. New York: Knopf, 1983.

Doyle, Arthur Conan. *History of Spiritualism*. 2 vols. New York: George H. Doran, 1926; single vol. ed., New York: Arno, 1975.

Edel, Leon, ed. *The Diary of Alice James*. New York: Dodd, Mead, 1964. Paperback, Baltimore: Penguin, 1974.

Fodor, Nandor. *An Encyclopedia of Psychic Science*. New Hyde Park, N.Y.: University Books, 1966. Paperback, New York: Citadel, 1974.

Gauld, Alan. "Mrs. Piper." In *Man, Myth, and Magic*, Vol. 16. New York: Marshall Cavendish, 1970.

Hansel, C. E. M. *ESP: A Scientific Evaluation*. New York: Scribner's, 1966. Revised and retitled, *The Search for Psychic Power*. Amherst, N.Y.: Prometheus Books, 1989.

James, William. "What Psychical Research Has Accomplished." In *The Will to Believe*. London: Longmans, Green, 1897.

———. "Final Impressions of a Psychical Researcher," and "Frederick Myers' Services to Psychology." In *Memories and Studies*. London: Longmans, Green, 1911.

Lodge, Oliver. *Past Years*. New York: Scribner's, 1931.

Murphy, Gardner, and Robert Ballou, eds. *William James on Psychical Research*. New York: Viking, 1960.

Piper, Alta L. *The Life and Work of Mrs. Piper*. Introduction by Oliver Lodge. London: Kegan Paul, 1929.

Podmore, Frank. *Mediums of the Nineteenth Century*, Vol. 2. New Hyde Park, N.Y.: University Books, 1963.

Rinn, Joseph. *Sixty Years of Psychical Research*. New York: Truth Seeker, 1950.

Tanner. Amy E. *Studies in Spiritism*. Introduction by G. Stanley Hall. New York: Appleton, 1910.

Yeazell, Ruth, ed. *The Death and Letters of Alice James*. Berkeley: University of California Press, 1981. Paperback, 1983.

31

"Dr." Henry Slade, American Medium

This essay first appeared in *The Encyclopedia of the Paranormal*, edited by Gordon Stein (Prometheus, 1996).

Nothing is known about Slade's private life—when or where he was born, who his parents were, or what education he had, if any. The "doctor" that he attached to his name had no factual support. He was described as six feet tall, slender and handsome, with dark eyes, a sad dreamy expression, silky gray hair, a small black mustache, and the ability, shared by all top psychic charlatans, to convey an impression of total sincerity. He always dressed elegantly and wore several diamond rings. William Eglinton, his successor in England, was more skillful at slate handling, but he never achieved Slade's notoriety.

Slade's specialty was the seemingly miraculous production, on slates previously examined by séance sitters, of banal messages purporting to come from discarnates. His techniques varied. A tiny piece of "slate pencil" (chalk) would be placed on a blank slate. While Slade held the slate beneath a table, scratching sounds would be heard, followed by three light raps to indicate the spirits had finished scribbling. Two slates, apparently shown blank on all four sides, would be tied together with a bit of pencil between them. Untied, a message would be revealed.

"Dr." Henry Slade

Occasionally a slate would mysteriously break in half. Why spirits would move chalk only when its movements were concealed from view was never explained.

Slade's spirit controls also produced many of the physical phenomena that had become the stock-in-trade of nineteenth-century mediums. Loud raps would occur during a séance. An accordion would be played by unseen hands. Chairs would slide and topple over. Tables would levitate. In semidarkness, sitters would be patted and pinched by invisible fingers.

In 1876, after fifteen years of fame in the United States, Slade visited London where his initial success was spectacular. Spiritualists hailed him as D. D. Home's worthy successor. Among the distinguished men of science and letters who declared him genuine were Lord Rayleigh, Sir William F. Barrett (who eulogized Slade in his 1918 book, *On the Threshold of the Unseen*), Alfred Russel Wallace, and Sir Arthur Conan Doyle.

The first heavy blow fell on Slade in 1876 when skeptic Sir Edwin Ray Lancaster, a zoologist, grabbed a slate during a séance, before the scratching began, and found a message already on it. After publishing an account of this in the *Times*, September 16, 1876, Lancaster took Slade to court for accepting money (one pound) under false pretenses. The judge's harsh sentence of three years at hard labor was overturned on a technicality. Before Lancaster could issue a second summons, Slade fled England.

In 1877, after holding séances in Holland, Denmark, and Germany, Slade went to Leipzig to be tested by Johann Carl Friedrich Zöllner, a respected but incredibly naive astronomer at the University of Leipzig. Zöllner had become obsessed by Kant's suggestion that there are higher Euclidean spaces as real as three-dimensional space. Zöllner believed these spaces were the abode of departed souls, and that mediums had the power to move objects in and out of the fourth dimension. To test his theory Zöllner devised a variety of simple experiments. He was assisted in this research by physicists Gustav Thedor Fechner, William Edward Weber, and others—all, like Zöllner, supremely ignorant of conjuring methods.

Slade succeeded only on tests that allowed easy trickery, such as producing knots in cords that had their ends tied together and the knot sealed, putting wooden rings on a table leg, and removing coins from sealed boxes. He failed utterly on tests that did not permit deception. He was unable to reverse the spirals of snail shells. He could not link two wooden rings, one of oak, the other of alder. He could not knot an endless ring cut from a bladder, or put a piece of candle inside a closed glass bulb. He failed to change the optical handedness of tartaric acid from dextro to levo. These tests would have been easy to pass if Slade's spirit controls had been able to take an object into the fourth dimension, then return it after making the required manipulations. Such successes would have created marvelous PPOs (permanent paranormal objects), difficult for skeptics to explain.

Zöllner wrote an entire book in praise of Slade. Titled *Transcendental Physics* (1878), it was partly translated into English in 1880 by Spiritualist Charles Carleton Massey. The book is a classic of childlike gullibility by a scientist incapable of devising adequate controls for testing paranormal powers.

In 1878 the French astronomer and science writer Camille Flammarion sealed two slates together with a piece of slate pencil between them in such a way that they could not be separated without this being detected. Slade kept the slates for ten days before he returned them as blank as before.

After his huge success in bamboozling poor Zöllner and conducting successful séances in Russia, Slade returned briefly to England where he tried to hide his identity by using the name "Dr. Wilson." Later séances in Australia are colorfully described by James Curtis in his three-hundred-page book *Rustlings in the Golden City: Being a Record of Spiritualistic Experiences in Ballarat and Melbourne* (London: Office of Light, 1902).

In 1885 Slade settled in New York City for the remainder of his life. That same year he was foolish enough to appear before the Seybert Commission, founded by Philadelphia Spiritualist Henry Seybert with funds given to the University of Pennsylvania's philosophy department. The commission's observers caught Slade in flagrant cheating. He was seen removing a foot from a slipper and using his toes to pinch and pat sitters. A supposedly blank slate, accidentally turned over, had writing on it. He was caught using a foot to shove empty chairs to or from a table, or to topple them over.

It was noticed that messages written on slates in response to questions asked during a séance were barely legible, whereas messages prepared in advance were always in clear handwriting. Members of the commission conjectured that when Slade held a blank slate below a table, the writing was produced by a thimble with a piece of slate pencil fastened to it. The device, known to magicians as a "nail writer," may have been attached to a "pull"—an elastic for drawing it up a sleeve. In his later years Slade may have become familiar with ingeniously made "flap slates" for producing chalked messages, and with the technique of writing with silver nitrate. Such writing is invisible, but turns white when sponged with salt water.

John W. Truesdale, in *Bottom Facts of Spiritualism* (1885), tells how he visited Slade without revealing his name, but slyly leaving in his overcoat pocket a letter addressed to Samuel Johnson. Slade's spirits also addressed him as Johnson. On a later visit, while Slade was out of the room, Truesdale found a slate with a message on it. He secretly added below: "Henry, watch out for this fellow; he is up to snuff—Alcinda." Alcinda was supposed to be Slade's dead wife, though there is no evidence he ever married. Truesdale records his extreme pleasure in witnessing Slade's surprise and discomfort when he saw the second message.

Joseph Rinn, in *Sixty Years of Psychical Research* (1950), writes of visiting Slade at 58 East Ninth Street in Manhattan. Slade was then living with a woman he called his niece, and who played the piano during séances. Spirits produced a slate message from Eva, whom Rinn said was his deceased sister, though actually she was still alive. Eva disclosed how "happy" she was on the other side. Rinn saw Slade scratching a slate with a fingernail to pretend writing was taking place on a previously prepared slate.

Slade's methods of slate writing were well known at the time to professional magicians. They are discussed at length in such early books as William E. Robinson's *Spirit Slate Writing and Kindred Phenomena* (New York: Munn & Co., 1898) (William Robinson performed magic under the name of Chung Ling Soo), Charles F. Pidgeon's [or possibly Elijah Farrington's] anonymous *The Revelations of a Spirit Medium* (St. Paul, Minn.: Farrington, 1891), David P. Abbott's *Behind the Scenes with the Mediums* (Chicago: Open Court, 1907), and Hereward Carrington's *The Physical Phenomena of Spiritualism* (1907). See also "Mr. Davey's Imitations by Conjuring of Phenomena Sometimes Attributed to Spirit Agency," by Richard Hodgson, in *Proceedings of the Society for Psychical Research* 8 (1892): 253–310. Harry Houdini was too young in Slade's day to have investigated him, but there is a chapter on Slade in Houdini's *A Magician Among the Spirits* (1924), followed by a chapter on slate-writing methods. A detailed, up-to-date account of slate-writing techniques are four volumes of *Master Slate Secrets* (1977), privately published for the magic trade by Al Mann, a New Jersey student of conjuring. The fourth volume includes a short biography of Slade and notes on his most important successors.

The Seybert Commission called Slade's methods "puerile in their simplicity." Houdini describes a séance with an engineer named R. K. Carter. Slade "dematerialized" a matchbox by placing it on a slate, holding the slate below the table, then taking it out with nothing on it. He put the slate back under the table, took it out, and the box was back on top. "This disappearance did not impress me greatly," said Carter, "as I concluded the whole secret of dematerialization consisted of turning the slate over and holding the box in place by a finger, then after showing the surface empty, the slate was again turned over on being replaced under the table."

Secretly turning a slate was one of Slade's simplest dodges. When he used his thimble to write on a slate's underside, he would invert the slate before bringing it into view to give the impression that the writing was done on the side of the slate pressed against the table's underside.

Remiginus Weiss, a former medium who testified against Slade at the Seybert investigation, told Houdini of a séance he had with Slade in Philadelphia in 1882. During one episode Slade read aloud a question he said he had written on a slate when actually what he wrote was the answer. As he reached for a wet sponge at the center of the table, his other hand dropped the slate out of sight long enough for Slade to turn it over secretly. He then pretended to sponge away the question by wiping the slate's blank side. Under the table the slate was turned again, then brought out to reveal the chalked answer.

Weiss claimed that when he confronted Slade with detailed explanations of how all his feats had been done—accomplices had watched everything through peepholes—Slade broke down and signed a confession admitting he was trickster. Pale and perspiring, he pleaded for mercy, saying he had no other way to earn a livelihood. Weiss claims Slade fainted dead away.

The confession is printed in Houdini's book. Why had Weiss never publicized it? Because, he said, he felt pity for Slade's physical condition. There is no evidence that Weiss's account of this is genuine. The "confession," if it ever existed, has not survived.

Although Slade's slate work was done in full light, more spectacular spirit manifestations required darkness, the séance room illuminated only by one candle or by low-burning gaslight. Slade often pretended to go into trances, during which he would moan and shudder, and from which he would emerge with impressive spasms and head jerks that made cracking noises in his neck.

At times Slade obviously used secret assistants. Zöllner describes a séance during which a bed behind a screen was seen to move. Later the screen fell over and broke in two pieces. "It was not our intent to do harm," the spirits explained on a slate. "Forgive what has happened." In his two-volume *History of Spiritualism* (1926) Doyle freely admits that Slade cheated, by only now and then. He considers Slade a medium of great powers, and takes seriously Curtis's report of a séance by Slade in Australia in which a beautiful woman is materialized in a dimly lit room, her eyes veiled. Sitters could hear her white silk dress rustle as she moved about the room.

Like Home, Slade surely had a fake hand that in semidarkness he could put on the table beside a real hand, freeing his other hand to produce such phenomena as ringing a bell under a table and flinging it across the room. He liked to hold an accordion at the end opposite the keys and bounce the other end up and down. One could hear the spirits play simple tunes such as "Home Sweet Home"—tunes easily played on the single octave of a tiny harmonica concealed in the mouth behind a mustache.

One of Slade's favorite ploys was to toss something—such as a piece of wood or coal, or a pocket knife—upward behind his back or when no one was looking, so that it seemed to drop from the ceiling. Zöllner reports drops of water falling from above. He actually assumed they came from the fourth dimension, never suspecting Slade of using a water pistol or flicking water with his fingers. On papers covered with soot Slade would produce in the darkness a print of a bare foot, or of two feet. He probably used a cardboard cutout that could be turned either way.

Few great mediums were caught cheating more often than Slade. In "A Survey of American Slate-Writing Mediumship" by Franklin Prince, a Reverend Stanley LeFevre Krebs used a ridiculously simple device for catching Slade in fraud. It was a small mirror on his lap. For misdirection Slade handed Krebs a slate to examine, but Krebs only pretended to examine it while he secretly looked down at his lap mirror. He saw Slade reach one arm below the table's edge and quickly exchange two slates for two others in a stack on the floor behind a window curtain in the back of his chair. (For biographical information on Krebs, and titles of his books, see *Who Was Who in America*.)

Slade pretended to move a rotating steel needle by waving a finger over it. In his lap mirror, Reverend Krebs saw Slade raise a foot to touch his shoe to the table's underside each time the needle moved. Having a magnet concealed in the toe of a shoe, or taped to a knee, is a well-known method of secretly influencing compass needles. Investigators untrained in magic frequently rule out the use of a concealed magnet on the grounds that they checked a psychic's hands and found them empty!

Sitters reported seeing Slade produce loud raps by banging his heel on a chair rung. The British magician John Maskelyne examined a table once used by Slade and found a loose wedge that could be used for making loud raps. On one occasion a sitter kicked a spirit "hand" that touched him under the table, causing Slade obvious pain. He was often caught switching slates by numerous techniques such as the one observed by Krebs, or by accidentally dropping a slate, then picking up another one.

In 1886 Slade was arrested for fraud in Weston, Virginia. As reported in the *Boston Herald* (February 2, 1886), skeptics outside a room secretly observed a séance through a crack under a door. They saw Slade move tables and chairs with his feet. One of his favorite tricks was to use a foot to kick over a book placed on the table so it projected over an edge.

Slade became an alcoholic in his declining years. The *New York World* (March 10, 1892) reported him seriously ailing in Jackson, Michigan. Here is the account in full:

IS DR. SLADE, THE MEDIUM, A WOMAN?
He Told His Doctors So, but He Explains that
He Is a Liar When Sick

Jackson, Mich., March 99.—Dr. Henry Slade, the famous spiritualist and clairvoyant, who is sick here, is said to be a woman. A well-known physician of this place told this to *The World* correspondent today:

"I have seen Slade twice during the last few days. He seemed averse to my placing my hands on him and frequently said: 'You will not betray me, doctor?' He was finally so averse to receiving my attentions that I left him and another doctor was called. In fact, several have been called in one way or another.

"The doctor who took my place came to me two days ago in great excitement, saying Slade was a woman and suffering with female trouble; that he had cried, admitted his sex and told a strange story. I told him to calm himself, observe further and report to me. He paid several visits subsequently and reported the most astounding things."

Another doctor said: "I was called later than Dr. ——— and suspected something from the symptoms. I told Slade of my suspicions and he admitted that he was a woman, begging me not to betray him. I refused to treat him."

"How did Slade account for his mustache, doctor?," was asked.

"He said he forced it to grow, and was growing hair on his head the same way."

The World correspondent interviewed Slade at his room in the hotel.

"I might say such things when I am sick," said Slade. "In fact, I lie like the devil when I am ill."

"But the doctors say on their own responsibility you are a woman."

"They are knaves to say anything about me when I am sick."

"Did you tell Dr. ——— that you are the mother of a child now in Amsterdam, Europe?"

"If I did it was a lie like the rest of what I said. I say I am a ——— liar when I am sick."

Dr. Slade says he is forty-seven years old. He began his performances as medium in this city thirty-two years ago, and was then very nearly, if not quite, thirty. He had a beardless, very pale face, wore his hair long and dressed in most remarkable style.

He said he had been accused before this of being a woman, and that he had a special doctor in New York to whom he always applied if he could get him.

Note that if Slade was forty-seven in 1882 he could not have been almost thirty in 1850. If Rinn is accurate, Slade died in September 1905, penniless and mentally ill, in a Belding, Michigan, hospital. According to Rinn he was eighty. Rinn bases his information on an obituary, but he does not say where it appeared. It was not in the *New York Times* or in the *Belding Banner*. Betty Jenkins, a Belding librarian, tells me that no Henry Slade is listed in the burial records from Ionia County.

Spinney Sanitarium, in Smyrna, about five miles from Belding, was torn down in 1924. The only other possibility, writes Ms. Jenkins, is the state asylum that once existed in Ionia, fifteen miles away. Exactly when and where Slade died, his age, and where he was buried remains unknown. When Rinn called the hospital to verify the newspaper obituary, a doctor told him that Slade was a hermaphrodite.

Slade obviously was a mountebank of limited talents as a magician, but skillful enough to deceive thousands of true believers in Spiritualism. One can only marvel that intelligent men such as Doyle and Zöllner could suppose that spirits behind the veil have nothing better to do than propel tiny bits of chalk over a slate, held where no one can see the action, to convey inane messages to loved ones. Such shenanigans are no longer practiced by mediums and psychics for the simple reason that methods of cheating are now well known and easily detected.

References

Curtis, James. *Rustlings in the Golden City*. London: Office of Light, 1902.

Doyle, Conan. *History of Spiritualism*. London: George H. Doran, 1926.

Houdini, Harry. *A Magician Among the Spirits*. New York: Harper Brothers, 1924.

Mann, Al . *Master Slate Secrets*. 4 vols. Freehold, N.J.: Al Mann Exclusives, 1977.

Prince, Walter F. "A Survey of Slate Writing Mediumship," *Proceedings of the American Society for Psychical Research* 15 (1921): 315–595.

Rinn, Joseph. *Sixty Years of Psychical Research*. New York: Truth Seeker, 1950.

Truesdale, John W. *The Bottom Facts Concerning the Science of Spiritualism*. New York: G.W. Dillingham, 1892.

Zöllner, Johann C. F. *Transcendental Physics*. Translated from the German by Charles Massey. Boston: Banner of Light, 1901.

Index

Page numbers in *italics* refer to illustrations.

About the Author

Born in 1914, Martin Gardner obtained a bachelor's degree in philosophy at the University of Chicago in 1936. Early jobs included reporter on the *Tulsa Tribune*, writer in the University of Chicago's press relations office, and case worker in Chicago's Black Belt. After four years as a Navy yeoman he began freelancing with sales of fiction to *Esquire*. In New York City he worked for eight years as contributing editor of *Humpty Dumpty's Magazine* before he began a twenty-five-year period with *Scientific American* as writer of the Mathematical Games column.

Gardner is the author of more than seventy books about mathematics, science, philosophy, and literature. He has written two novels, *The Flight of Peter Fromm* and *Visitors from Oz*, and a collection of short stories, *The No-Sided Professor*. His two *Annotated Alice* books have been combined in a single volume published in 1999 by W. W. Norton. Among his awards are two honorary doctorates and several prizes for science and math writing. His main hobby is magic, about which he has written several technical books.

Gardner calls himself a philosophical theist in the tradition of Plato, Kant, Pierre Bayle, Charles Peirce, William James, and Miguel de Unamuno.